# teaching elementary school mathematics through motor learning

# teaching elementary school mathematics through motor learning


*By*

**ROBERT B. ASHLOCK**

*Professor and Director of the Arithmetic Center*
*College of Education*
*University of Maryland*
*College Park, Maryland*

*and*

**JAMES H. HUMPHREY**

*Professor and Motor Activity Learning Specialist*
*College of Physical Education, Recreation and Health*
*University of Maryland*
*College Park, Maryland*

**CHARLES C THOMAS • PUBLISHER**

***Springfield • Illinois • U.S.A.***

*Published and Distributed Throughout the World by*

CHARLES C THOMAS • PUBLISHER
Bannerstone House
301-327 East Lawrence Avenue, Springfield, Illinois, U.S.A.



ISBN 0-398-03578-4

Library of Congress Catalog Card Number: 76-13387

*With* THOMAS BOOKS *careful attention is given to all details of manufacturing and design. It is the Publisher's desire to present books that are satisfactory as to their physical qualities and artistic possibilities and appropriate for their particular use.* THOMAS BOOKS *will be true to those laws of quality that assure a good name and good will.*

*Printed in the United States of America*
*R-11*

*Library of Congress Cataloging in Publication Data*

Ashlock, Robert B 1930-
Teaching elementary school mathematics through motor learning.

1. Mathematics--Study and teaching (Elementary)
2. Perceptual-motor learning. I. Humphrey, James Harry, 1911- joint author. II. Title.
QA135.5.A784 372.7 76-13387
ISBN 0-398-03578-4

# PREFACE

CHILD learning through motor activity has received a great deal of recognition in recent years, and educators have been directing an increasing amount of attention to this social-psychological phenomenon. The use of motor activity as a medium of learning is not only supported by theoretical postulation but is backed up by research as well. *Teaching Elementary School Mathematics Through Motor Learning* has been developed with this in mind.

The introductory chapter gives an overview of the development of the teaching of mathematics in elementary schools over the years and points to the use of motor activities as a part of instructional programs in mathematics. Chapter 2 explains the theory of motor activity learning, while Chapter 3 suggests some general ways in which learning experiences in mathematics can be provided through motor activity. Chapter 4 takes into account some objective evidence for this approach as it applies to elementary school mathematics. Ways of integrating mathematics and reading through mathematics motor activity stories are suggested in Chapter 5. Chapters 6 through 8 describe a large number of motor activities in which a broad range of mathematical concepts are inherent.

A book of this type will have a variety of uses. It can be used as a supplementary text in teacher preparation courses in elementary school mathematics, and also as a handbook of mathematics learning activities for special education and classroom teachers.

Most of the materials contained in the text have undergone extensive field trials in a variety of situations. The authors are grateful to the many teachers who tested the materials and made valuable suggestions for their use.

R. B. A.
J. H. H.

*College Park, Maryland*

# CONTENTS

Page

*Preface* .......... v

*Chapter*

1. Mathematics in Today's Elementary Schools .......... 3
2. The Nature of Learning through Motor Activity .......... 20
3. General Ways of Providing Mathematics Experiences through Motor Activity .......... 38
4. Research in Learning about Mathematics through Motor Activity .......... 53
5. Mathematics Motor Activity Stories .......... 68
6. Learning about Number and Numeration Systems through Motor Activity .......... 82
7. Learning about the Operations of Arithmetic through Motor Activity .......... 105
8. Learning about Other Areas of Mathematics through Motor Activity .......... 130

*Appendices*

A. List of Activities by Concept .......... 147
B. List of Activities by Title .......... 153

*Index* .......... 157

# teaching elementary school mathematics through motor learning

Chapter 1

# MATHEMATICS IN TODAY'S ELEMENTARY SCHOOLS

TEACHING mathematics in today's elementary schools, with the never-ending innovations in method and changes in content, is understandably bewildering to the conscientious teacher who must weather the bombardment of new teaching methods — to specify competencies, develop mathematics laboratories, and make increasing use of games in instruction. This is not the first era of change in our schools, however. There was a time when mathematics was not even considered a proper subject of study for children. This chapter will present an historical perspective of this area of education.

## The Colonial Period

During the colonial period, the ability to compute was regarded as appropriate for a person doing menial work, but such skill was not viewed as appropriate for the aristocracy. Accordingly, the study of mathematics was not emphasized in colonial schools, not even the study of arithmetic.[1] Gradually, as commerce increased, the ability to compute became increasingly valued, and arithmetic became a part of the general education of the young; it gained an equal place in the curriculum with religion, reading, and writing.[2] In 1789, laws to make arithmetic a mandatory school subject were passed in Massachusetts and New Hampshire.[3] During the colonial period, arithmetic was used primarily by businessmen and very gradually came into the schools of the day. By 1800, arithmetic was taught quite generally in the schools.

---

[1]Monroe, Walter S., "Development of Arithmetic as a School Subject," Bulletin No. 10. (Washington, Bureau of Education, Department of the Interior, 1917), p. 5.

[2]DeVault, M. Vere, and Kriewall, Thomas E., *Perspectives in Elementary School Mathematics.* (Columbus, Charles E. Merrill Pub. Co., 1969), p. 4.

[3]Monroe, *Development of Arithmetic,* p. 13.

Arithmetic, as taught during the early decades of the new nation, consisted of "working problems from rules." Only the teacher had a book, and the rules presented were applied largely to problems of commerce of that day. Arithmetic was seldom taught to children below ten years of age. In fact, when a boy started to study the subject, it was considered a sign of approaching manhood.[4]

## The Nineteenth Century

The first two decades of the nineteenth century are sometimes regarded as the lowest point in the history of American schools, particularly with the introduction of the Lancasterian system around 1812. In this system teachers taught very large groups of children with the help of older children serving as monitors.

However, about 1820 the character of instructional practice began to change. This was largely due to the influence of Pestalozzi, a Swiss philosopher whose ideas were enjoying great popularity in Europe and whose writings were beginning to appear in America.

Among the first texts to reflect the ideas of Pestalozzi were those authored by Warren Colburn. His text published in 1821 and entitled *First Lessons in Arithmetic on the Plan of Pestalozzi* gradually became the text *par excellence* for teaching arithmetic to children. Colburn attempted to use an approach which was more inductive, he sequenced topics better, and instruction in arithmetic soon began at a much earlier age. Concerning this text, DeVault and Kriewall write:

> The book stressed the development of mental power through an understanding of the rationale of arithmetic, a practice as dramatic and challenging then as teaching the structure of mathematics is viewed in many quarters today . . . Three out of four of the first thousand exercises were for drill! However, the striking feature of the text lay in its attempt to introduce arithmetic inductively to children in the context of socially meaningful

---

[4]DeVault and Kriewall, *Perspectives,* p. 46.

> illustrations and materials. Postponement of drill was advocated until an understanding of the basic processes had been developed.[5]

Colburn introduced two major changes: instruction sequenced from the concrete to the abstract, and large numbers of drill exercises.

Another follower of Pestalozzi was A. W. Grube, a German whose texts became popular in parts of the United States in the middle 1800s. He and his followers advocated an inductive approach which made use of objects. They included all four operations (addition, subtraction, multiplication, and division) in work with small numbers (1-10) before working with larger numbers.[6]

As reformers, Colburn and Grube exerted less influence than might have been expected. Most of the arithmetic text writers and classroom teachers of the last half of the nineteenth century were influenced heavily by the faculty psychology of the phrenologists who believed the mind was composed of no less than thirty-seven "faculties." These faculties included memory, reasoning, will, and the like. Each faculty was thought to be a kind of mental muscle for which continued exercise was needed if it was to develop.[7] Joseph Ray's texts, which were very popular during this period, were based upon faculty psychology, as were the texts by Daniel Fish. In his suggestions for teachers in an 1880 arithmetic, Fish writes, "The teaching of arithmetic must, therefore, to a great extent, be considered as disciplinary, — as training and developing certain faculties of the mind, and thus enabling it to perform its functions with accuracy and dispatch."[8]

It is not surprising that mathematics was viewed by large numbers of children as something to be dreaded, for exercises were deliberately designed to be difficult in order to better "exercise" the mind. One page of Fish's 1871 text has addition examples,

---

[5]DeVault and Kriewall, *Perspectives*, p. 7.

[6]Riedesel, C. Alan, *Guiding Discovery in Elementary School Mathematics*, 2nd ed. (New York, Appleton, 1973), p. 17.

[7]DeVault and Kriewall, *Perspectives*, p. 50.

[8]Fish, Daniel W., *The Complete Arithmetic*. (New York, Ivison, Blakeman, Taylor & Co., 1880), p. vi.

each with 25 five- and six-digit addends![9] Examples of written problems provided for children include the following:

> "What is the cost of 9¼ tons of coal, if .875 of a ton cost $5.635?"[10]
>
> "Bought 6/7 of a box of candles, and having used 7/8 of them, sold the remainder for 16/25 of a dollar; how much would a box cost at the same rate?"[11]

It was thought that such problems would help children think clearly and quickly. It is questionable whether today's instruction in mathematics has as yet completely recovered from the accompanying dread of arithmetic and the idea that "difficult is good" and "fun must be bad."

As schools became graded during the late nineteenth century, texts also became graded. Whereas a single text had been sufficient, texts were written for each grade level. The first such series of texts appeared in 1877. They were written by Joseph Ray and were oriented toward faculty psychology. Graded texts resulted in more specific expectations at each level, locking teachers and children into material they were expected to cover. By the close of the nineteenth century, arithmetic instruction had become an "empty formalism."[12]

However, there were many seeking changes by the latter part of the century. The thinking of philosophers such as Herbart and James began to influence pedagogical thinking. The value of a formal discipline approach, based upon faculty psychology, was being questioned. Rather, attention was turning toward practical values. The Child Study Movement was also beginning to have its impact. It is not surprising that the period from 1890 to 1911 has been thought of as a period of reflection in mathematics education.[13]

### The Twentieth Century

Several different movements have already left their mark on

---

[9]Fish, Daniel W., *Robinson's Progressive Practical Arithmetic.* (New York, Ivison, Blakeman, Taylor, & Co.), 1871, p. 27.
[10]Fish, *Robinson's Progressive,* p. 180.
[11]Fish, *Robinson's Progressive,* p. 115.
[12]DeVault and Kriewall, *Perspectives,* p. 10.
[13]Wilson, Guy M., Stone, Mildred B., and Dalrymple, Charles O., *Teaching the New Arithmetic.* (New York, McGraw, 1939), p. 46.

mathematics education during the twentieth century, some running concurrently and others as a reaction to a different point of view. Teachers in today's schools, educated in different places and at different times, frequently continue to reflect the emphases of these various movements.

### *Stimulus-Response Theories*

Early in the century, scientism came to have great influence in education, largely through the work of Edward Thorndike. His research and thought were in a large measure responsible for ridding elementary schools of faculty psychology as a dominant force.[14] It was Thorndike's conclusion that learning involved forming "bonds" or connections, links between a specific stimulus and a particular response. His arithmetic texts, therefore, stressed the introduction of very specific skills according to a hierarchical scheme. Such bonds were to be strengthened by extensive drill.

Problems were selected that were better suited to the social needs of the adult population and did not include the absurd problems found in texts of previous decades. On the other hand, students were expected to approximate 100 percent accuracy. Drill was no longer valued for "training the brain," but it was valued as the means of establishing essential bonds. The following quotation from a Thorndike arithmetic text summarizes his thinking:

> Reasoning is treated, not as a mythical faculty which may be called on to override or veto habits, but as the cooperation, organization and management of habits ... Nothing that is desirable for the education of children in quantitative thinking is omitted merely because it is hard, but the irrelevant linguistic difficulties, the unrealizable pretenses at deductive reasoning, and the unorganized computation which have burdened courses in arithmetic are omitted. The demand here is that pupils shall approximate 100 percent efficiency with thinking of

---

[14]DeVault and Kriewall, *Perspectives*, p. 14.

which they are capable.[15]

Arithmetic may not have involved mental gymnastics, but it was still heavily drill-oriented. Often it made little sense to the pupil. It is not surprising then that children frequently disliked the subject.

### *Social Utility Theory*

Though Thorndike restricted the difficulty of his selection of problems when compared with educators influenced by faculty psychology, the effort to restrict problems in elementary arithmetic to those encountered in the normal daily lives of the adult population is associated with the work of Guy Wilson. In the preface of his text on methods of teaching arithmetic, Wilson stated that the "school should no doubt develop arithmetic somewhat beyond the present actual needs of children, but certainly not beyond the needs of adults. Common adult usage should be the limit of any arithmetic undertaken for drill mastery in the grades and general high school."[16] The influence of social utility theory was greatly extended as research on the actual use of arithmetic began. In 1939, Wilson stated:

> The present period is marked as beginning in 1911, for it was in that year that a new force, the survey of usage, began to operate. The studies on usage were reported in statistical form rather than as mere argument and thus became more powerful in influencing practice. Processes having little or no practical value were definitely listed for elimination.[17]

Some of the advocates of social utility theory believed that with the program simplified, results should be much better and 100 percent mastery of the identified "fundamentals" should be expected.

The emphasis on social utility resulted in the development of problem units for the different grades. For example, "A grocery store at school" in grade two, "The home garden, does it pay?" in

---

[15]Thorndike, Edward L., *The Thorndike Arithmetics: Book Two.* (Chicago, Rand, 1924), p. v.

[16]Wilson, Stone, and Dalrymple, *Teaching the New*, p. vii.

[17]Wilson, Stone, and Dalrymple, *Teaching the New*, p. 47.

grade four, and "The family budget" in grade six. One of Wilson's suggestions for grade three was "Games with teams."[18]

For the first two grades, Wilson emphasized the social development of the child and his progress in reading. He stated:

> Into this program of the first two years, arithmetic enters therefore only in connection with games and meaningful activity, the purpose being to make the child at home with simple numbers and to lead him to use numbers exactly as he uses language. If there is sufficient repetition in a game, it would appear that the children will know the number facts . . . But the mastery of number facts is not the chief aim.[19]

Wilson appreciated the value of games for instruction, but did not want to see games used for drill or practice purposes until grade three. Concerning the first two grades, he stated:

> Games should be used here to give number meaning and experience. The child playing beanbags, with three numbers upon which his bags may fall, will have opportunity to read and write these numbers and to add any two of them . . . He may not immediately associate the numbers added with their sum, but when he does he will know the meaning of the primary facts that he has learned in a useful setting. Facts so learned will not be just the repetition of syllables.[20]

Children studying arithmetic under the influence of social utility theory were not usually without drill exercises in the course of their progress through the grades, but arithmetic *was* understood to be something which would be used, and instruction often involved more informal pupil participation. Games were encouraged.

## Gestalt Theories and Piaget

By 1930, elementary school mathematics curricula were increasingly influenced by Gestalt or field theories of learning which were articulated most notably by Max Wertheimer in

---

[18]Wilson, Stone, and Dalrymple, *Teaching the New*, pp. 309-317.
[19]Wilson, Stone, and Dalrymple, *Teaching the New*, pp. 104-105.
[20]Wilson, Stone, and Dalrymple, *Teaching the New*, p. 79.

Germany. According to Gestalt theories, learning was not a matter of establishing connections. Instead, learning involved the identification of meaningful patterns within an individual's field of perception. As a result, instruction was organized to enable the learner to focus on the whole rather than very small parts. Drill came to be emphasized less for it was thought that mastery required less drill if learning occurred in a meaningful context.

Contemporary with the increasing influence of Gestalt psychology was a greater awareness of the research and thinking of the Swiss psychologist, Jean Piaget. As his writings were translated and became more generally available, an awareness of a child's stages of intellectual development added significantly to the increased popularity of experiences with number which involved "concrete" materials. More recently, the work of Jerome Bruner in the United States has widely influenced elementary school mathematics programs along lines consistent with Gestalt theories and in general agreement with Piaget's ideas.

### *Meaning Theory*

The influence of Gestalt or field theories of learning upon mathematics instruction was initially seen in a shift from stressing only social meanings toward more of a stress on mathematical meanings. The term *meaning theory* is commonly associated with this movement which had considerable impact on instruction in the 1940s and 1950s. In *The Teaching of Arithmetic,* a 1935 yearbook of the National Council of Teachers of Mathematics, William Brownell stated that meaning theory "conceives of arithmetic as a closely knit system of understandable ideas, principles, and processes."[21] Stress on socially useful arithmetic had too often been accompanied by rote instruction on the fundamentals and by drill which made little mathematical sense to the child. Educators advocating meaning theory stressed the need for helping children *understand* mathematical

---

[21]Brownell, William A., Psychological considerations in the learning and the teaching of arithmetic, *The Teaching of Arithmetic,* Tenth yearbook of the National Council of Teachers of Mathematics, (Washington, 1935), p. 19.

principles and processes, and they taught that drill was to be used only to reinforce material the child already understands.

Over the years, the word *meaningful* has been used with so many different referents that confusion has understandably developed. When the term is used to refer to instruction in which the mathematics makes sense to the child, e.g. he understands *why* he renames in addition computation, then it is appropriate to contrast meaningful instruction with rote instruction. However, since the 1930's meaningful instruction has often been contrasted with drill or practice. As a result, dangerously little practice has been included in some programs.[22]

## CONTEMPORARY MATHEMATICS PROGRAMS

Changes in elementary school mathematics programs since the mid-1950's have been rather dramatic. These changes can be viewed as an acceleration of the changes toward more mathematically meaningful instruction which had taken place during the previous two decades, perhaps with a change of focus. Several factors converged to help bring about the "revolution" which occurred.

### Contributing Factors

Mathematics itself had changed, and attempts to unify mathematical concepts led to new basic structures which had not yet been reflected in mathematics instruction below the university level.* Mathematics had come to be based largely upon set theory and logic. Another contributing factor was the accumulating information about how children learn, for it was becoming well established that children *could* learn quite complex concepts,

---

[22]DeVault and Kriewall, *Perspectives,* p. 18.

*For background on the changes in mathematics which finally affected the curriculum in mid-century, see Wilder, R. L., Historical background of innovations in mathematics curricula, *Mathematics Education,* Sixty-ninth yearbook of the National Society for the Study of Education, Part I, (Chicago, University of Chicago Press, 1970), pp. 7-22.

often at a younger age. Other factors often cited include the concern that the mathematics curriculum was largely the result of historical development rather than a logical entity with unifying themes, the increasing need for an understanding of mathematics by people in business and industry, and a conviction on the part of many that there was an overemphasis on computational skills.

## Characteristics of New Programs

The elementary school mathematics programs that were developed during the late 1950's and the 1960's focused heavily upon concepts and principles — especially structures which could help unify the curriculum. In 1963, Edwina Deans noted: "Experimental groups throughout the country today are trying to identify the unifying ideas pervading all mathematics and to determine ways of beginning the development of these ideas in the elementary school."[23] This emphasis was reflected by the National Council of Teachers of Mathematics in its 1959 yearbook titled *The Growth of Mathematical Ideas: Grades K-12.*[24] The content of elementary programs characteristically involved more algebraic ideas and more geometry than had been included in previous years. Relationships between operations were stressed, e.g. the inverse relationship between addition and subtraction. Properties of such operations came to be a major focus of instruction, ideas like commutativity, associativity, and identity elements. Familiar topics were also extended, e.g. work in numeration was extended to bases other than ten.

Mathematics educators also sought to change the method of instruction so that learning would be characterized by discovering patterns and testing hypotheses. Many of the patterns to be discovered by each child were the very structures considered to be unifying themes — properties of operations, numeration patterns, and so forth. However, whereas content could be changed

---

[23]Deans, Edwina, *Elementary School Mathematics: New Directions.* U. S. Dept. of Health, Education, and Welfare Bulletin No. 13 (OE-29042) (Washington, U. S. Government Printing Office, 1963), p. 3.

[24]National Council of Teachers of Mathematics, *The Growth of Mathematical Ideas: Grades K-12.* Twenty-fourth yearbook, (Washington, 1959).

in large part through new textbooks and curriculum guides, the task of orienting teachers to a new method of instruction proved to be more difficult. Often, the new content was presented in old ways.

## Criticisms of the New Programs

Few teachers would choose to return to the elementary school arithmetic programs which existed before the advent of "new math," but some of the contemporary programs have been strongly criticized by the general public and by educators alike. Gradually, evidence has been collected to suggest that children have not done as well with the most formal and rigorous textbooks as they have done with more informal and less rigorous programs.[25]

## Structural Apparatus

Increasing emphasis upon structure and unifying ideas can be understood as a direct outgrowth of Gestalt psychology. Recall that in Gestalt theories, learning is considered to be a matter of observing patterns or structures within the pupil's field of perception, i.e. within his sensory experiences. Gestalt or field theories of learning, along with the research and writings of Piaget, have prompted the development of structural apparatus. These are concrete aids for learning which illustrate mathematical structures by their very design, yet can be handled and moved about in the solution of problems. In recent years such materials have come into rather general use as elementary schools have begun to move toward more informal mathematics programs.

Maria Montessori was developing such apparatus in Italy early in the century, and her materials continue to be used in special schools. Catherine Stern's structural materials for teaching arithmetic, developed during the 1940s, have not been used as universally as have the rods developed by the Belgian, Georges Cuisenaire. More recently, Zolten P. Dienes, presently at

---

[25]Glennon, Vincent J., Current status of the new math, *Educational Leadership*, XXX, pp. 604-607.

Sherbrooke University in Canada, has developed both logical and multibase blocks which he advocates using in game-like activities designed to help the child abstract patterns or mathematical structures inductively.

## The Laboratory Approach

During the 1970's, an increasing number of teachers have come to use some form of the laboratory approach for teaching elementary school mathematics. Some writers have gone so far as to call the movement "a new revolution," believing that the reforms of the 1960's have lost their momentum.[26] They feel that the laboratory approach constitutes an improvement in method, whereas the efforts of the previous decade only accomplished reforms in content.

The term *mathematics laboratory* is difficult to define, because many people have interpreted it differently. However, Reys and Post believe it has at least two distinct connotations: "One is that of an approach to learning mathematics, while the other is that of a place where students can be involved in learning mathematics. These notions encompass the physical aspects of a room with materials and the educational purposes which the laboratory is designed to achieve."[27] A mathematics laboratory has many materials, and probably different kinds of structural apparatus are included. But the critical thing is that children use these materials while *actively* involved in learning activities, activities which are often of a problem-solving nature. Accordingly, the term "active learning" is frequently used in literature concerned with the laboratory approach. Investigations concerning the environment are often encouraged, and it is believed that they "will do far more good than traditional teaching methods to build enthusiasm for and confidence in mathematics."[28]

---

[26]Kidd, Kenneth P., Myers, Shirley S., and Cilley, David M., *The Laboratory Approach to Mathematics.* (Chicago, Science Research Associates, Inc., 1970), pp. xi-xii.
[27]Reys, Robert E., and Post, Thomas R., *The Mathematics Laboratory: Theory to Practice.* (Boston, Prindle, 1973), p. 9.
[28]Kidd, Myers, and Cilley, *The Laboratory,* p. 10.

Though a laboratory approach to elementary school mathematics can be identified with some of the reform efforts in the United States, e.g. the Elementary Science Study and some of the activities of the Madison Project, we recognize that during the 1970's British experience with the laboratory approach has had considerable impact in fostering a similar approach among schools in the United States. Most notable have been the activities and publications of the Nuffield Project* and the writings of Edith E. Biggs.[29] In addition, the research and writings of Piaget have had an especially great influence on the movement toward developing mathematics laboratories.

With much emphasis upon active involvement of children in learning situations, it is not surprising that there has come to be increased emphasis upon the use of games in instructional settings.

## GAMES FOR LEARNING

Mathematics educators are increasingly concerned with the attitudes of children toward mathematics and with the general problem of motivation. Some writers, such as John Biggs, have concluded that "anxiety appears to be more easily aroused in learning mathematics than it is in other subjects."[30] In recent years, the instructional possibilities of games have come to be recognized increasingly.

### A Viable Alternative

Biggs and MacLean, in their book on the laboratory approach,

---

*Publications of the Nuffield Mathematics Project are available in the United States from John Wiley and Sons, Inc., 605 Third Ave., New York, New York 10016.

[29]See especially Biggs, Edith E., and MacLean, James R., *Freedom to Learn: An Active Learning Approach to Mathematics*. (Don Mills, Ontario, Canada, Addison-Wesley Ltd., 1969).

[30]Biggs, John: The psychopathology of arithmetic, in F. W. Land, ed., *New Approaches to Mathematics Teaching*, (London, Macmillan & Co., Ltd., 1963), p. 59.

note that not only are games fun, but they are also "tremendously useful devices for developing skill in mathematics. Practice in computational skills is just as effective and much more palatable when disguised in a game context."[31] Their focus is on interest, and one advantage of using games for instruction *is* increased attending behavior. Kennedy and Michon note, as do other mathematics educators, that games can be used in the development of concepts as well as the mastery of skills.[32] Games can also help individualize instruction. They are valued by mathematics educators for variety and novelty, while at the same time providing needed structure. Repeated exposure to a concept or skill is inherent in many game situations. Kennedy and Michon summarize the advantages of games as follows:

> Games help teachers overcome problems connected with how children learn mathematics. They give children variety in the way they deal with a topic, allow them to actively participate in the learning process, provide repeated exposures without becoming tiresome, and enrich children's backgrounds.[33]

Ashlock notes two characteristics of games which make them especially valuable for helping children master the basic facts of arithmetic.[34] In the first place, a child does not *have* to win a game every time, whereas it is the ultimate expectation of much instruction that the child will get the correct answer every time. In a game, the child feels more free to be wrong; he is not humiliated if he does not win. Of course he must win some of the time, and when children have needed prerequisite skills and are grouped carefully, they *will* win part of the time. Secondly, if a child is to master the basic facts of arithmetic he must practice "pulling them out of his head" instead of always figuring them out the long way. A game situation frequently provides the prompting to respond quickly which is needed if recall is to be reinforced.

---

[31]Biggs and MacLean, *Freedom to Learn,* p. 50.

[32]Kennedy, Leonard M., and Michon, Ruth L., *Games for Individualizing Mathematics Learning.* (Columbus, Merrill, 1973), p. v.

[33]Kennedy and Michon, *Games for Individualizing,* p. 2.

[34]Ashlock, Robert B., *Error Patterns in Computation.* (Columbus, Merrill, 1972), pp. 6-8.

## The Slow Learner

Whereas much of the research and writing during the 1960's focused upon what the majority of children can do, by the end of the decade there was renewed concern for the slow learner, the reluctant learner, the disadvantaged child. Mathematics educators, in addressing the needs of the slow learner, have given considerable importance to the use of games. In the 1972 yearbook of the National Council of Teachers of Mathematics, *The Slow Learner in Mathematics*, repeated reference is made to games.[35] In fact, the first appendix actually describes many games.[36] Pearson notes that the slow learner in mathematics typically needs a different kind of environment than the usual textbooks and worksheets.

> Coming from an environment that values the quick hand and strong, fast body, the slow learner is confused when placed in the school environment, which values the clear head and glib tongue. Although the thoughtful, verbal style is important, it is much better to capitalize on the motor skills and watch the other develop in time.[37]

Games that involve the child physically are viewed as especially necessary when teaching the slow learner. Pearson states that, for the sake of variety, many of these activities can be conducted outdoors.[38] In this regard it will be noted in Chapter 4 that our research in physically-oriented learning experiences suggests that this approach is highly favorable for children with below-average intelligence.

Schultz, in describing the characteristics and needs of the slow learner, stresses the need for alternatives, novelty, and variety in

[35]National Council of Teachers of Mathematics, *The Slow Learner in Mathematics*, Thirty-fifth yearbook, (Washington, 1972).

[36]Rowan, Thomas E., and McKenzie, William G.: Appendix A: activities, games, and applications, *The Slow Learner in Mathematics*, Thirty-fifth yearbook of the National Council of Teachers of Mathematics, (Washington, 1972), pp. 444-486.

[37]Pearson, James R., A favorable learning environment, *The Slow Learner in Mathematics*, Thirty-fifth yearbook of the National Council of Teachers of Mathematics, (Washington, 1972), p. 114.

[38]Pearson, *The Slow Learner*, p. 115.

mathematics instruction.[39] He states that slow learners demand novelty and variety which provide alternatives to accommodate their many learning styles, and he believes that games involving mathematics are an important source of that variety and novelty.

Concern with the education of slow learners has also centered on the need for person-centered activities which involve direct action and visible results. For this type of activity, Paschal recommends games.[40] Riessman concurs, and states:

> . . . most games . . . are person-centered and generally are concerned with direct action and visible results. Games are usually sharply defined and structured, with clear-cut goals. The rules are definite and can be readily absorbed. The deprived child enjoys the challenge of the games and feels he can 'do' it; this is in sharp contrast to many verbal tasks.[41]

Action and visible results are also emphasized in the observation of Marks, Purdy, and Kinney that games and relay races are particularly effective with slower pupils,[42] and in Junge's recommendation for the use of marching activities when establishing readiness for multiplication and division.[43]

When describing the need for games and physical activities in the teaching of mathematics to slow learners, mathematics educators repeatedly acknowledge the value of these same activities for other children as well. The commercial market has taken advantage of the increased interest in games and a great variety of games has become available. While many of the commercial games help the child sustain interest in the subject and involve some physical motion, most do not involve the kind of extensive bodily move-

---

[39]Schultz, Richard W., Characteristics and needs of the slow learner, *The Slow Learner in Mathematics,* Thirty-fifth yearbook of the National Council of Teachers of Mathematics, (Washington, 1972), pp. 15-16.

[40]Paschal, Billy J., Teaching the culturally disadvantaged child, *The Arithmetic Teacher,* XIII, p. 372.

[41]Riessman, Frank: *The Culturally Deprived Child.* (New York, Har-Row, 1962), p. 71.

[42]Marks, John L., Purdy, C. Richard, and Kinney, Lucien B., *Teaching Elementary School Mathematics for Understanding,* 3rd ed. New York, McGraw-Hill Book Co., 1970, p. 342.

[43]Junge, Charlotte W., Adjustment of instruction (elementary school), *The Slow Learner in Mathematics,* Thirty-fifth yearbook of the National Council of Teachers of Mathematics, (Washington, 1972), p. 138.

ment that we associate with learning through motor activity. The nature of learning through such activities is the topic of the next chapter.

Chapter 2

# THE NATURE OF LEARNING THROUGH MOTOR ACTIVITY

## Meaning of Motor Learning

THE many definitions for the term *motor learning* tend to revolve around the general theory that motor learning is concerned only with the learning of motor skills. As an example, the *Dictionary of Education* states that ideational learning is concerned with ideas, concepts, and mental associations, while motor learning is that in which the learner achieves new facility in the performance of bodily movements as a result of specific practice.[1] Although this may be a convenient and simple description of motor learning, it does not serve the purpose adequately in modern times. The reason for this is that motor learning can no longer be considered a unilateral entity. At one time, when thought of only in terms of learning motor skills, it might have been considered by some as almost the exclusive purview of the physical educator and psychologist. However, it has now become such a multiphasic area that it compels the interest and attention of a variety of professions and disciplines.

Various aspects of motor learning in one way or another are involved in such fields as physical education, psychology, psychiatry, and neurophysiology. In fact, in almost any endeavor of human concern that one might mention, some aspect of motor learning could play a very significant part. It is for this reason that motor learning is no longer thought of only in the sense of the previously mentioned definition. Consequently, some attempt needs to be made to identify certain branches of specific aspects of motor learning. This is the case with this volume.

---

[1]Good, Carter V., *Dictionary of Education,* 2nd ed. (New York, McGraw, 1959), p. 314.

## Branches of Motor Learning

In order to put this particular topic into its proper perspective, and to establish a suitable frame of reference for subsequent discussions of it, the authors will identify three specific aspects of motor learning.[2] It should be understood that these identifications are used arbitrarily for the authors' own purpose. Others may identify these differently, and in the absence of anything resembling standardized terminology, it is their prerogative to do so. Moreover, it is also recognized that some individuals might wish to segment these aspects of motor learning further, or add others. With this in mind, the authors will identify the three aforementioned aspects as follows:

1. motor learning which is concerned essentially with conditions surrounding the *learning of motor skills,*
2. motor learning which is concerned essentially with *perceptual motor development* (This may also be referred to as *psychomotor development* or *neuromotor perceptual training.*), and
3. motor learning which is concerned essentially with *academic skill and concept development.*

It should be clearly understood that all of the above areas of motor learning involve the same general concept, and that there are various degrees of interrelatedness and interdependence among areas.

### *Motor Learning Involving the Learning of Motor Skills*

This branch of motor learning has commanded the attention of individuals in the field of physical education mainly because it forms the basic citadel for subject matter and methods of teaching in this field. Some of the areas in which attention has been centered include: how individuals learn motor skills, length and distribution of practice, mechanical principles, transfer, and

[2]To our knowledge this particular classification of motor learning was first introduced into the literature as follows: Humphrey, James H., Academic skill and concept development through motor activity. *The Academy Papers,* No. 1, pp. 29-35, 1968.

retention. Research in some of these areas by some physical educators has been outstanding and, as might be expected, it has been done primarily by those who have some background in psychology.

In recent years it appears that there has been more compatibility between psychologists and physical educators regarding this branch of motor learning. The fact that such has not always been the case is suggested in the following comment by one psychologist, "There is one perhaps distressing feature which is apparent: this is the seeming lack of awareness which the two disciplines have of the progress *and* problems of the other's area."[3]

Fortunately there is evidence of amelioration of this condition because psychologists are more frequently discovering that physical education and sports experiences provide an excellent natural climate and laboratory for the study of human performance and behavior. The authors should mention again that this aspect was the only subject of definition of motor learning in the *Dictionary of Education*. Moreover, it will *not* be the function of the present text to deal with this branch of motor learning as such. There are a number of fine books devoted entirely to this branch of motor learning, and which may be found in most major libraries.

### *Motor Learning Involving Perceptual Motor Development*

This branch of motor learning involves the correction, or at least some degree of improvement, of certain motor deficiencies, especially those associated with fine coordinations. The need for this type of training is exemplified in certain neurologically handicapped children who may have various types of learning disorders. What some specialists have identified as a "perceptual-motor deficit" syndrome is said to exist in such cases. An attempt may be made to correct or improve fine motor control problems through a carefully developed sequence of motor competencies, which follow a definite hierarchy of development. This may

---

[3]Johnson, G. B., Motor Learning. In W. R. Johnson, ed., *Science and Medicine of Exercise and Sports*. (New York, Harper and Brothers, 1960), p. 602.

occur through either structured or unstructured programs. As far as physical education is concerned, this branch of motor learning might be thought of as education *of* the physical.

In recent years, various aspects of this area of motor learning have contributed to the alleviation of certain types of learning disorders in children. It should be apparent that a wide range of specialists and disciplines are needed to help children effectively through this area of motor learning. The authors previously called attention to certain areas of speciality which in one way or another can make important contributions. Again, it will *not* be the function of this volume to explore this dimension of motor learning. However, there is a rather substantial body of literature listed in the various educational indices which the interested reader may pursue in the area of perceptual-motor development.

### *Motor Learning Involving Academic Skill and Concept Development Through Motor Activity*

This might also be referred to as the "physical education learning medium." This branch of motor learning is concerned specifically with children learning basic skills and concepts in the various subject areas in the elementary school curriculum through the medium of motor activity. Whereas motor learning involving perceptual-motor development was thought of in terms of education *of* the physical, this aspect of motor learning is concerned with education *through* the physical. The focus and thrust of this book will be on the contribution this aspect of motor learning can make to child learning of skills and concepts in mathematics.

It may be interesting to note that one well-known psychologist, Dr. James J. Asher, has identified this branch of motor learning as "total physical response motor learning."[4] Asher suggests that research in motor learning is usually done with tasks which involve *parts* of the body — the receptors and effectors — as illustrated in display-control problems, rather than with the whole

---

[4]Asher, James J., "The total physical response technique of learning," *Journal of Special Education,* Fall, 1969.

body. His work with the total physical response technique has been done in the area of foreign language education, and essentially involves having students listen to a command in a foreign language and immediately respond with the appropriate physical action. In connection with our work, Asher comments as follows:

> The work that comes closest to total physical response motor learning is Humphrey's investigations of learning through games in which, for example, the acquisition of certain reading skills was significantly accelerated when the learning task occurred in the context of a game involving the entire body.[5]

## THE THEORY OF LEARNING THROUGH MOTOR ACTIVITY

The idea of motor activity learning is not new. In fact, the application of motor activity was a basic principle of the Froebelian kindergarten and was based on the theory that children learn and acquire information, understanding, and skills through motor activities in which they are naturally interested, such as building, constructing, modeling, painting, and various forms of *movement.*

Movement is one of the most fundamental characteristics of life. Whatever else they may involve, all of man's achievements are based upon his ability to move. Obviously, the very young child is not able to perform abstract thinking, and he only gradually acquires the ability to deal with symbols and intellectualize his experiences in the course of his development as Piaget has so ably documented. On the other hand, the child is a creature of movement and feeling. Any efforts to educate the child must take this relative dominance of the intellectual versus movement and feeling into account. Furthermore, by engaging in movement experiences to develop academic skills and concepts, these skills and concepts become a part of the child's physical reality.

When the authors speak of motor activity learning, they refer to things that children *do* actively in a pleasurable situation in order to learn. This should suggest to teachers that motor learning

---

[5]Asher, *Journal of Special.*

activities can be derived from basic physical education curriculum content found in such broad categories as *game* activities, *rhythmic* activities, and *self-testing* activities, hence the authors' reason for indicating previously that learning through motor activity can also be referred to as the physical education learning medium. Classroom teachers and physical education teachers might well team up to facilitate learning and development of skills in their respective areas of responsibility through an integrated and cross-reinforcement program.

This brings up an important point for consideration, that is, the function of the classroom teacher and physical education teacher in the use and implementation of motor activity learning. It has been mentioned before that motor activity learning experiences can be derived from physical education content. However, physical education should be considered a subject "in its own right" in the elementary school curriculum with the natural corollaries of good teaching, sufficient facilities, etc. Consequently, the use of motor activity as a learning medium in other subject areas — in this case mathematics — should not ordinarily occur during the regular time allotted to physical education. On the other hand, this approach should be considered as a way of learning mathematics in the same way that other kinds of learning activities are used in this subject. This means that when this procedure is used, it should be employed during the regularly allotted time for mathematics, with the ideal situation allowing the classroom and physical education teachers to work closely together in the use of this approach. The classroom teacher knows the mathematics skills and concepts to be developed and the physical education teacher should have a thorough knowledge of the various physical education activities that might be useful in developing these skills and concepts.

This aspect of motor learning is based essentially on the theory that children — being predominantly movement-oriented — will learn better when what the authors will arbitrarily call "academic learning" takes place through pleasurable physical activity, that is, when the *motor* component operates at a maximum level in skill and concept development in school subject areas essentially oriented to so-called "verbal" learning. This is *not* to say that "motor" and "verbal" learning are two mutually exclusive kinds of learning, although it has been suggested that at the two extremes the dichotomy appears justifiable. It is recognized that verbal learning, which involves almost complete abstract symbolic manipulations, may include such motor components as tension, subvocal speech, and physiological changes in metabolism which operate at a minimal level. It is also recognized that in physical education activities where the learning is predominantly motor in nature, verbal learning is evident, although perhaps at a minimal level. For example, in teaching a physical education activity there is a certain amount of verbalization in developing a kinesthetic concept of the particular activity that is being taught.

One way to use motor activity learning involves the selection of a physical education activity which is taught to the children and used as a learning activity for the development of a skill or concept in a specific subject area — in the case of the present text, the subject area of mathematics. An attempt is made to arrange an *active* learning situation so that a fundamental intellectual skill or concept is being acted out, practiced, or rehearsed in the course of participating in the activity.

Essentially, there are two general types of such activities. One type is useful for developing a specific concept, where the learner "acts out" the concept and thus is able to visualize as well as to get the "feel" of the concept. Concepts become a part of the child's physical reality as the child participates in the activity where the concept is inherent.

For example, a concept to be developed may be *multiples of three.* An activity in which this concept is inherent is a game called *Pick up Race.* In this game, several wooden blocks (three for each player) are scattered over a large playing area. The players, divided into a number of teams, take places behind a starting line and near the circle which has been assigned to their team. On a signal from the leader, each player runs into the playing area, picks up one block, returns to the starting line and places it in his team's circle. He goes back after the second block, and then again until he has brought back his third block and placed it in a pile within the circle. Play continues in this manner until all blocks have been picked up and placed in a circle in groups of three. The team which finishes bringing back the blocks first to the circle is declared the winner.

In one specific situation where this activity was used the class counted out how many threes would be needed for each team. By counting they found that if there would be ten on a team they would need thirty blocks. The blocks were placed in groups of three so the children who were having difficulty could see the groups of three. The number of players was changed frequently so the groups of three would change. Many of them could see where the multiples of five and two could be learned through this game. In this particular situation this game proved to be a very good way of teaching multiples of three. One child who could not understand that four groups of three would be twelve and five groups of three would be fifteen, etc. soon developed an understanding of the threes. As the number of players on the team were changed, the children counted how many threes would be within the circle when the team finished play.

The second type of activity helps to develop skills by using these skills in highly interesting and stimulating situations. Repetitive drill for the enrichment of skills related to specific concepts can be utilized.

For example, an activity called *Call and Catch* can be used for practicing the *ordinal use of numbers.* The children stand in a circle, the teacher also stands in the circle, holding a rubber ball. Each child is assigned a different number. The teacher throws the ball into the air and calls out a number by saying, "Just before six" or "Next after five." For example, if the teacher calls "Next after five," the child assigned number six tries to catch the ball after it bounces. The teacher can provide for individual differences of children. For example, for the slower child the teacher can call the number and then momentarily hold the ball before throwing it into the air. Experience has shown that this activity can provide children the opportunity to become familiar with the sequence of numbers as they practice counting forward and backward.

## FACTORS INFLUENCING LEARNING THROUGH MOTOR ACTIVITY

During the early school years, and at ages six to eight

particularly, it is possible that learning is limited frequently by a relatively short attention span rather than by intellectual capabilities alone. Moreover, some children who do not appear to think or learn well in abstract terms can more readily grasp concepts when given an opportunity to use them in an applied manner. In view of the fact that the child is a creature of movement and also that he is likely to understand concrete terms rather than abstract terms, it would seem to follow naturally that the motor activity medium would suit him well.

The above statement should not be interpreted to mean that the authors are suggesting that movement-oriented experiences (motor learning) and passive learning experiences (verbal learning) are two different kinds of learning. The position is taken here that learning is learning, even though, as stated previously, in the motor activity approach the motor component may be operating at a higher level than in most of the traditional types of learning activities.

The theory of learning accepted here is that learning takes place in terms of reorganization of the systems of perception into a functional and integrated whole because of the result of certain stimuli. This implies that problem solving is the way of human learning and that learning takes place through problem solving. If a motor activity learning situation is to be well planned, a great deal of consideration should be given to the inherent possibilities for learning in terms of problem solving. In this approach opportunities abound for near-ideal teaching-learning situations because of the many problems to be solved. In playing active games, the following representative questions asked by children indicate that there is a great opportunity for reflective thinking, use of judgment, and problem solving in this type of experience.

1. Why didn't I get to touch the ball more often?
2. How can we make it a better game?
3. Would two circles be better than one?
4. What were some of the things you liked about the game?
5. How can I learn to throw the ball better?

Another very important factor to consider with respect to this approach is that a considerable part of the learnings of young children are motor in character, with the child devoting a good proportion of his attention to skills of a locomotor nature. Furthermore, learnings of a motor nature tend to usurp a large amount of the young child's time and energy and are often closely associated with other learnings. In addition, it is well known by experienced classroom teachers at the primary grade levels that the child's motor mechanism is active to the extent that it is almost impossible for him to remain for a very long period of time in a quiet state as may be required by the passiveness of the learning situation.

Demanding of children prolonged sedentary states is actually, in a sense, in defiance of a basic physiological principle which is concerned directly with the child's basic metabolism. The term *metabolism* is concerned with physical and chemical changes in the body which involve producing and consuming energy. The rate at which these physical and chemical processes are carried on when the individual is in a state of rest represents his *basal metabolism*. Thus, the basal metabolic rate is indicative of the speed at which body fuel is changed to energy, as well as how fast this energy is used.

Basal metabolic rate can be measured in terms of calories per unit of body surface, with a calorie representing a unit measure of heat energy in food. It has been found that on the average, basal metabolism rises from birth to about two or three years of age, at which time it starts to decline until between the ages of 20 to 24. Also the rate is higher for boys than for girls. With the relatively highest metabolic rate and therefore the greatest amount of energy occurring during the early school years, deep consideration might well be given to learning activities through which this energy can be profitably utilized. Moreover, it has been observed that there is an increased attention span of primary-age children during active play. When a task such as a motor learning experience is meaningful to a child, he can spend longer periods engaged in it than is likely to be the case in some of the more traditional types of learning activities.

The comments made thus far have alluded to some of the general aspects of the value of the motor activity learning medium. The ensuing discussions will focus more specifically upon what we call certain *inherent facilitative factors* in this approach which are highly compatible with child learning. These factors are motivation, proprioception, and reinforcement, all of which are somewhat interdependent and interrelated.

## Motivation

When considering motivation as an inherent facilitative factor of learning through motor activity, the authors would like to think of the term as it is described in the *Dictionary of Education,* that is, "the practical art of applying incentives and arousing interest for the purpose of causing a pupil to perform in a desired way."[6]

We need also to take into account *extrinsic* and *intrinsic* motivation. Extrinsic motivation is described as "appreciation of incentives that are external to a given activity to make work palatable and to facilitate performance," while intrinsic motivation is the "determination of behavior that is resident within an activity and that sustains it, as with autonomous acts and interests."[7]

Extrinsic motivation has been and continues to be used as a means of spurring individuals to achievement. This most often takes the form of various kinds of reward incentives. The main objection to this type of motivation is that it tends to focus the learner's attention upon the reward rather than the learning task and the total learning situation.

In general, the child is motivated when he discovers what seems to him to be a suitable reason for engaging in a certain activity. The most valid reason of course is that he sees a purpose for the activity and derives enjoyment from it. The child must feel that what he is doing is important and purposeful. When this occurs and the child gets the impression that he is being successful in a group situation, the motivation is intrinsic, since it comes about naturally as a result of the child's interest in the activity. It is the premise here that the motor activity learning medium contains

---

[6]Good, *Dictionary,* p. 354.
[7]Good, *Dictionary,* p. 354.

this "built-in" ingredient so necessary to desirable and worthwhile learning.

The ensuing discussions of this section of the chapter will be concerned with three aspects of motivation that are considered to be inherent in the motor activity learning medium. These are (1) motivation in relation to *interest,* (2) motivation in relation to *knowledge of results,* and (3) motivation in relation to *competition.*

### *Motivation in Relation to Interest*

It is important to have an understanding of the meaning of interest as well as an appreciation of how interests function as an adjunct to learning. As far as the meaning of the term is concerned, the following description given some time ago by Lee and Lee expresses in a relatively simple manner what is meant by the terms *interest* and *interests*: "Interest is a state of being, a way of reacting to a certain situation. Interests are those fields or areas to which a child reacts with interest consistently over an extended period of time."[8]

A good condition for learning is a situation in which a child agrees with the learnings which he considers of most value to him, i.e. those that are of greatest interest to him. To the very large majority of children, their active play experiences are of the greatest personal value to them.

Under most circumstances, a very high interest level is concomitant with pleasurable physical activities simply because of the expectations of pleasure children tend to associate with such activities. The structure of a learning activity is directly related to the length of time the learning act can be tolerated by the learner without loss of interest. Motor learning experiences, by their very nature, are more likely than many of the traditional learning activities to be so structured.

### *Motivation in Relation to Knowledge of Results*

Knowledge of results is most commonly referred to as *feedback.*

---

[8]Lee, J. Murray and Lee, Dorris May, *The Child and His Development,* New York, (Appleton, 1958), p. 382.

It was suggested by Brown many years ago that feedback is the process of providing the learner with information concerning the accuracy of his reactions.[9] Ammons has referred to feedback as knowledge of various kinds which the performer received about his performance.[10]

It has been reported by Bilodeau and Bilodeau that knowledge of results is the strongest, most important variable controlling performance and learning, and further that studies have repeatedly shown that there is no improvement without it, progressive improvement with it, and deterioration after its withdrawal.[11] As a matter of fact, there appears to be a sufficient abundance of objective evidence that indicates that learning is usually more effective when one receives some immediate information on how he is progressing. It would appear rather obvious that such knowledge of results is an important adjunct to learning because without it one would have little idea which of his responses was correct.

The motor activity learning medium provides almost instantaneous knowledge of results because the child can actually *see* and *feel* himself throw a ball, or tag, or be tagged in a game. He does not become the victim of a feedback mechanism in the form of a poorly constructed paper and pencil test, the results of which may have little or no meaning for him.

### *Motivation in Relation to Competition*

Using games as an example to discuss the motivational factor of competition, we will describe games as *active interactions of children in cooperative and/or competitive situations*. It is possible to have both cooperation *and* competition functioning at the same time, as in the case of team games. While one team is

---

[9]Brown, J. S., "A proposed program of research on psychological feedback (knowledge of results) in the performance of psychomotor tasks." Research Planning Conference on Perceptual and Motor Skills, AFHRRC Conference Report, 1949, U. S. Air Force, San Antonio, Texas, pp. 1-98.

[10]Ammons, R. B., "Effects of knowledge of performance: A survey and tentative formulation," *Journal of General Psychology, 54*: 279-99, 1956.

[11]Bilodeau, Edward A., and Bilodeau, Ina, "Motor skill learning," *Annual Review of Psychology*, Palo Alto, California, pp. 243-270, 1961.

competing against the other, there is cooperation within each group. In this framework it could be said that a child is learning to cooperate while competing. It is also possible to have one group competing against another without cooperation within the group, as in the case of games where all children run for a goal line independently and on their own.

The terms *cooperation* and *competition* are antonymous; therefore, the reconciliation of children's competitive needs and cooperative needs is not an easy matter. In a sense, we are confronted with an ambivalent condition, which, if not carefully handled, could place children in a state of conflict.

Modern society not only rewards one kind of behavior (cooperation), but also its direct opposition (competition). Perhaps more often than not our cultural demands sanction these rewards without providing clear-cut standards regarding the specific conditions under which these forms of behavior might well be practiced. Hence, the child is placed in somewhat of a quandary when deciding when to compete and when to cooperate.

As far as the competitive aspects of some motor activities such as active games are concerned, they not only appear to be a good medium for learning because of their intrinsic motivation, but this medium of learning can also satisfy the competitive needs of children in a pleasurable and enjoyable way.

## Proprioception

Earlier in this chapter, the accepted theory of learning was stated to the effect that learning takes place when systems of perception are reorganized into a functional and integrated whole as a result of certain stimuli. These systems of perception, or sensory processes as they are sometimes referred to, ordinarily consist of the senses of sight, hearing, touch, smell, and taste. Armington has suggested that "although this point of view is convenient for some purposes, it greatly oversimplifies the ways by which information can be fed into the human organism."[12] He indicates that

[12]Armington, John C., *Physiological Basis of Psychology*, (Dubuque, Iowa, Wm. C. Brown Co., Publisher, 1966), p. 16.

a number of sources of sensory input are overlooked, particularly the senses that enable the body to maintain its correct posture. As a matter of fact, the 60 to 70 pounds of muscle which include over 600 in number that are attached to the skeleton of the averaged-sized man could well be his most important sense organ.

Various estimates indicate that three fourths of our knowledge is derived from our visual sense. Therefore, it could be said with little reservation that man is *eye minded.* However, Steinhaus has reported that "a larger portion of the nervous system is devoted to receiving and integrating sensory input originating in the muscles and joint structures than is devoted to the eye and ear combined."[13] In view of this, Steinhaus has contended that man is *muscle-sense minded.*

Generally speaking, *proprioception* is concerned with muscle sense. The proprioceptors are sensory nerve terminals that give information concerning movements and position of the body. A proprioceptive feedback mechanism is involved which in a sense regulates movement. In view of the fact that children are so movement oriented, it appears reasonable to speculate that proprioceptive feedback from the receptors of muscles, skin, and joints contributes in a facilitative manner when the motor activity learning medium is used to develop mathematic skills and concepts. The combination of the psychological factor of motivation and the physiological factor of proprioception inherent in the motor activity learning medium has caused us to coin the term *motor*v*ation* to describe this phenomenon.

## Reinforcement

If the motor activity learning medium is to be considered compatible with reinforcement theory, the meaning of reinforcement needs to be taken into account. *Reinforcement* can be acceptably described as an increase in the efficiency of a response to a stimulus brought about by the concurrent action of another stimulus. The motor activity learning medium reinforces attention

[13]Steinhaus, Arthur H, Your muscles see more than your eyes, *Journal of Health, Physical Education and Recreation,* September, 1966.

to the learning task and learning behavior, and keeps children involved in the learning activity, which is perhaps the major area of application for reinforcement procedures. Moreover, there is perhaps little in the way of human behavior that is not reinforced or at least reinforcible by feedback of some sort, and the importance of proprioceptive feedback has already been discussed in this particular connection.

In summarizing this discussion, it would appear that the motor activity learning medium generally establishes a more effective situation for learning reinforcement for the following reasons:

1. The greater motivation of the children in the motor activity learning situation involves accentuation of those behaviors directly pertinent to their learning activities, making these salient for the purpose of reinforcement.
2. The proprioceptive emphasis in the motor activity learning medium involves a greater number of *responses* associated with and conditioned to learning stimuli.
3. The gratifying aspects of the motor activity learning situations provide a generalized situation of *reinforcers.*

## CURRENT STATUS AND FUTURE PROSPECTS OF THE MOTOR ACTIVITY LEARNING MEDIUM

One of the reasons given for studying the history of a subject is that it helps to determine how the past has challenged the present, so that we might better understand how the present might challenge the future. As far back as the early Greeks, it was suggested by Plato that learning takes place better through play and play situations. Similar pronouncements over the years include those made by Aristotle, Quintillian, Rousseau, Froebel, and Dewey. Despite many esteemed endorsements of the motor activity approach to learning, there is not a great deal of historical evidence that indicates that there was widespread practice of it in the schools. As mentioned previously, Friedrich Froebel, who is considered the founder of the kindergarten, incorporated the practice as a part of the school day. Perhaps, having been influenced by some of his predecessors, John Dewey indicated his thoughts on the matter in 1919 with reference to this approach to learning in the actual school situation. He commented that "Experience has shown that when children have a chance at physical activities which bring their natural impulses into play, going to school is a joy, management is less of a burden, and learning is easier."[14]

In more recent times the use of motor activity learning in education has been given considerable attention. A concrete example of this is a statement made in the publication *The Shape of Education for 1966-67.* A chapter entitled "Learning is Child's Play," makes reference to games in education at the University of Maryland, where "a whole series of playground games have been devised for teaching the elements of language, science, arithmetic and such matters to elementary school children." Further, "The strange thing about all these incidents is that they are isolated and unrelated but that they demonstrate an educational trend for which there is respectable theoretical justification in serious academic research."[15]

It seems worthy of mention that in contemporary society there is a certain degree of universality in the use of motor activity for learning. While the ideologies of the eastern and western worlds may differ in many respects, apparently there is some agreement about this approach to learning. For example:

> There was even an arithmetic game in which there were two teams of fifteen players each and an umpire. Ten students on each side would represent a number from one to ten, and the other five players would represent the symbols used in arithmetic: plus (+); minus (-); division (÷); multiplication (x); and equals (=). The umpire would shout out an example such as two times four minus eight equals zero, and each side would rush to line up in this order. The team to do so first was the winner.[16]

In a somewhat similar frame of reference, the Russian Ivanitchkii, has stated, "Teachers of mathematics and physics

---

[14]*Dewey, John, Democracy in Education, An Introduction to the Philosophy of Education,* (New York, Macmillan, 1919), pp. 228-229.

[15]National School Public Relations Association, *EDUCATION U. S. A., The Shape of Education for 1966-67.* (Washington, D. C., 1966), p. 49.

[16]Hunter, Edward, *Brain-Washing in Red China.* (New York, Vanguard, 1951), p. 46.

should use examples of sports in solving problems."[17]

Up to this point, we have only been extolling the rectitude of the motor activity approach to learning. Certainly we would be remiss if we did not call attention to some of its possible limitations. However, the suggested limitations of it are likely to center around inertia of individuals and tradition rather than the validity of the medium itself. In any event, some people in education may feel that pupils will not take the approach seriously enough as a way of learning and therefore will not concentrate on the skill or concept being taught. However, our personal experience with this medium has been quite the contrary. In another sense, some may fear that this medium of learning may be *too* attractive to children. For example, in many of our experiments on this approach, children have asked, "Why don't we learn it this way all the time?"

It should be pointed out very forcefully here that we do not recommend that learning through motor activity be the tail that wags the "educational dog." But, rather we would like to look upon it as *another* valid way that children might learn and not necessarily the only way. We are well aware of the fact that everything cannot be taught best through motor activity.

Although it is difficult to predict what the future holds for this medium of learning, we feel that more serious attention is currently being paid to it. Discussions with leading neurophysiologists, learning theorists, child development specialists, and others reveal a positive attitude toward the motor activity learning medium. And there is rather general agreement that it is very sound from all standpoints: philosophical, physiological, and psychological.

---

[17]Ivanitchkii, M. F., "Physical education of school children — the constant concern of all pedagogical collectives." *Theory and Practice of Physical Culture* 4:10, 1962 (Russian). From an abstract by Michael Yessis *Research Quarterly*, 35:339, 1964.

Chapter 3

# GENERAL WAYS OF PROVIDING MATHEMATICS EXPERIENCES THROUGH MOTOR ACTIVITY

INHERENT in active play situations are many motor learning opportunities for such processes as counting, computing, and measuring. Although the main purpose of this text is to deal with more or less specific ways of learning mathematics through motor activity, some mention should be made about how this can occur generally in such broad categories of activities as *game* activities, *rhythmic* activities, and *self-testing* activities.

## Mathematics Learning Experiences in Game Activities

In certain types of games, such as tag games, the number of children caught can be counted. After determining the number caught, pupils can be asked how many were left. The number caught can be added to the number left to check for the correct answer. In this procedure, counting, adding, and subtracting are utilized as number experiences in the game activity. An example of this is shown in the game *Lions and Tigers*.

In this game the class is divided into two groups, the Lions and the Tigers. Each group stands on a goal line at opposite ends of the playing area. Both groups face in the same direction. To start the game the Lions come up behind the Tigers as quietly as possible. The Tigers listen for the Lions and when it is determined from the sound that they are near, one person who has been designated calls out, "The Lions are coming!" The Tigers turn and give chase and try to tag as many as they can before the Lions reach their own goal line. Those caught can become members of the opposite group. The above procedure is reversed and the game proceeds.

In one situation this game was used for a better understanding of *comparison of groups, and more or less.* The class was working on the one-to-one relationship between groups. This game gave the class another opportunity to use this concept in a real situation. The children understood that each team would try to add to their players. The winner was determined by counting the Lions and Tigers. There were 18 Lions and 12 Tigers. The Lions and Tigers stood in a line face to face to get another demonstration of the one-to-one concept. There were some Lions who did not have a Tiger to face. The class was able to see clearly that one group was larger than the other. This group had *more* children in it, and the other *less.* The class drew the following conclusions: The higher we count, the bigger the number. Therefore, 18 is more than 12; 12 is less than 18; 9 is more than 6; 6 is less than 9. The class made many such group comparisons.

In games that require scoring, for example, the number of runs scored in a baseball-type game, there are opportunities for counting, adding, and subtracting as well as for using fractions. Pupils can compute how many *more* runs one group scored and how many *fewer* runs another group scored. Teachers may find it useful to score with a number in which the class needs practice in adding. For example, a run could be worth five points and the number five would be added whenever someone scored. Also, one base may be counted as one-fourth of the way around the bases, two bases one-half of a run, and so on to demonstrate in a concrete way the use of fractions.

In games in which children have numbers and they go into action when their number is called, mathematics can be used by having the leader who calls the number give a problem which will have as its answer the number for the players who are to go into action. Any of the four operations of addition, subtraction, multiplication, or division, or any combination of these can be used.

An example of such a game is *Club Snatch.* It is played with two teams from 8 to 16 players on a team, forming two lines facing each other about 10 to 12 feet apart. A small object such as a beanbag or a tenpin (club) is placed in the middle of the space between the two lines. The team members are numbered from opposite directions. The teacher, or a pupil acting as the leader, calls a number and each of the two players with that number runs out and tries to grab the object and return to his line. If the player does so, his team scores two points; if he is tagged by the opponent, the other team scores one point. The teacher might give a problem in the following manner: "Ready! 10 minus 2 divided by 4 times 3 plus 1, Go!" The answer would be 7, and the two players having this number would attempt to retrieve the object. Care should be taken to have the competing children as nearly equal in ability as possible.

In some game activities various kinds of records can be kept to allow children to experience computations with numbers, although this does not involve direct participation. For example, in determining team or group standings on a won and lost basis, there is an opportunity for intermediate-level pupils to gain insight into decimals and percentages. Team standings can be determined by dividing the number of games won by the number of games played. If a group won six games and lost four out of a total of ten games played, six would be divided by ten, giving a percentage of .600.

Various game activities require boundaries with certain dimensions for the playing area. Pupils can measure and lay out the necessary boundaries. In studying mapmaking to scale, intermediate-level pupils can draw field dimensions for such activities and layout boundaries from their scale drawings.

Opportunities are available for use of mathematics experiences in the *organization* of game activities. The following sample questions are submitted to illustrate this procedure.

"If we divide our class into four teams, what part of the class would each team be?"

"We have 32 people present today. What number would we have to divide by to get eight in each group?"

## Mathematics Learning Experiences in Rhythmic Activities

There are many rhythmic activities in which mathematics experiences are inherent. A case in point is the singing game *Ten*

*Little Indians.* In this activity the children form a circle, all facing in. Ten children are selected to be Indians, and numbered from one to ten. As the song is sung, the child whose number is called skips to the center of the circle. When the ten little Indians are in the center, the song is reversed. Again, each child leaves the center and returns to the circle as his number is called. Other children may become Indians, and the song is repeated. The words of the song are:

One little, two little, three little Indians,
Four little, five little, six little Indians
Seven little, eight little, nine little Indians,
Ten little Indian boys (or girls).
Ten little, nine little, eight little Indians,
Seven little, six little, five little Indians,
Four little, three little, two little Indians,
One little Indian boy.

This activity can be used to develop the understanding of *numbers and counting*. In this rhythmic activity children can be helped to relate the cardinal and ordinal uses of numbers.

As in the case of game activities, organizing for rhythmic activities can provide for mathematics experiences. This is illustrated with the following sample questions:

"If each person has a partner in this dance, how many couples will there be in our room?"

"If we need eight or nine couples for our dance, how could we divide?"

"One couple is what part of a square dance set? Two couples? Three couples?"

## Mathematics Learning Experiences in Self-Testing Activities

Numbers can be incorporated satisfactorily into the teaching of certain self-testing activities in a variety of ways. For example, counting can be facilitated for young children by counting the number of times they bounce a ball. Addition and subtraction can be brought in here also by having them compute the number of times they bounced the ball one time and how many more or fewer times they bounced it another time. This same procedure can be applied to activities such as rope-jumping, as indicated by the following examples:

"Let the rope swing forward seven times before you jump over it."

"Joan, after the rope turns five times, you run in and jump nine times."

"Susan jumped four times. How many more times must she jump to make ten?"

Because so many stunt activities can be broken down into a number of parts, there is an opportunity to teach about fractions in connection with this kind of activity. An example is the *Squat Thrust* which is performed in the following manner: From a standing position the child assumes a squatting stance, placing the hands to the surface area to the outside of the legs with the palms flat and the fingers pointed forward. This is the first count. On the second count the weight is shifted to the hands and arms, and the legs are extended sharply to the rear until the body is straight. The weight of the body is now on the hands and the balls of the feet. On the third count the child returns to the squatting position, and on the fourth count the child returns to the erect standing position.

In teaching this stunt with reference to fractional parts of a whole (1/4, 2/4, 3/4) as well as addition of fractions, the following auditory input might be furnished in connection with the performance of the stunt.

> Here is a stunt with four parts. First you stand straight with feet together. Hands are at your side. Next you stoop down to a squat position. Your hands are in front on the floor. Now you have done the first part of the stunt. You have done 1/4 of the stunt. Next you kick your legs way back. Now you have done 2/4 of the stunt. Next you bring your legs back to the squat position. Now you have done 3/4 of the stunt. Next you stand up straight again. Now you have done the whole stunt.

As additional input, the teacher can make a circle out of cardboard and cut it into 1/4's. As the child does each part of the stunt the 1/4's can be put together. Another variation would be to draw a circle on the floor. The child makes his movements in the circle, calling out the fractional parts as he does so.

In keeping records of various kinds of self-testing activities, practical use can be made of graphs and charts to show pupil progress. Each pupil can graph or chart his progress in any activity that requires the use of numbers. An example of this procedure would be in keeping records of individual pupils in such skills as throwing for accuracy. A graph or chart could be made over a period of several days indicating how well he throws for accuracy at a target.

## Simulated Teaching-Learning Situations

Over a period of time a large number of teaching-learning situations involving the learning of mathematics through motor activity have been developed. This has been done by tape recording dialogue between teachers and children in given situations. This material can be used to great advantage by teachers as an evaluative technique to help determine the success of the teaching-learning situation. The following are some representative examples of such teaching-learning situations.

The first situation involves a procedure which integrates a rhythmic activity with the study of fractions. The name of the activity is *Pop Goes the Weasel* and this version of it is carried out in the following manner.

Groups of three, with hands joined, form a small circle. These groups are arranged in a large circle as in the following diagram:

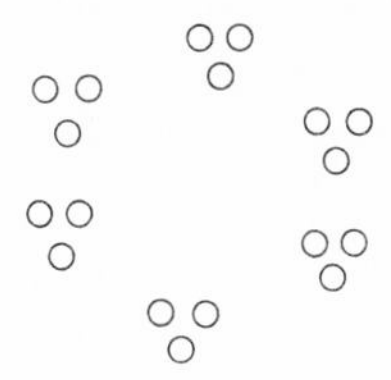

Each group walks around counterclockwise until the strain "Pop Goes the Weasel" is heard on the recording. At this time one pupil is "popped" under the joined hands of the other two and sent to the next couple in the large circle. Progression of those popped is counterclockwise. Before starting the dance, it must be decided which person in each group will be "popped" through, and the sequence in which the other two will be "popped" through.

*Teacher*: Boys and girls when reading about sports events in the paper the other day, I came across some sentences that I thought would be of interest to you. In a story about a football game, one sentence said, "The halfback danced down the field." In another report there was a sentence that said, "The runner danced back and forth off of first base." As I thought about this, it occurred to me that dancing might be a good way to develop some of the skills needed to play well in sports and games. Have any of you ever heard of how rhythm and dancing could help to make better players?

*Pupil*: Gee, I never thought of it that way before.

*Pupil*: Me, either.

*Teacher*: Please form a large circle with a boy and a girl in every other place. *(Pupils form the circle)* Today we are going to learn a dance called "Pop Goes the Weasel." Probably many of you have heard the song before. Let me see the hands of those who have. (*Some pupils indicate that they are familiar with the song.*) Do you remember the other day, when we were studying fractions, what the bottom number told us?

*Pupil*: I think it is how many parts something is divided into.

*Teacher*: Yes, that's right, and now we are going to divide our large circle into groups of three. Starting here, we will form small circles with three persons in each small circle with hands joined.

(*Teacher demonstrates with first group of three to her left.*)

Now each group is a circle made up of three persons. If something is divided into three parts, what do we call each part?

*Pupil*: One-third?

*Teacher*: Yes, that's right. Can someone tell me how many thirds make a whole?

*Pupil*: Three.

*Pupil*: (*Aside to another pupil*) Oh! I see what she meant the other day.

*Teacher*: All right, now let's have each person in the small circles take the name of either First-Third, Second-Third, or Last-Third. Just take a minute to decide within your own circle who will take each part.

(*The teacher demonstrates with one of the small circles*)

Let me see the hands of all the First-Thirds, the Second-Thirds, the Last-Thirds. Now here is something that is very important. The magic word in this dance is "pop." (That is, if the words to the song are to be sung). We will move to the right in our circles. When we come to the part of the music where the word "pop" would be sung, the person in each circle with the name First-Third will pop under the arms of the other two persons in his circle and become a part of the circle to his right. The people in the small circles will immediately begin to walk again, and this time on the word "pop" the person named Second-Third will leave his circle. The next time the person named Last-Third leaves. Who do you think will pop the next time?

*Pupil*: Would we start all over again with First-Third?

*Teacher*: Yes, that's right, and we will continue that way until the record is finished playing. I am going to play a part of the record for you. This time I want you to listen to the music, and when you hear the part where you pop into the next circle raise your hand so that we can make sure that you will know when to pop ... That's fine. Everyone seems to know when to pop. Listen for the chord to start.

(*Pupils participate in the dance, and after the record is played through once the teacher evaluates the activity with them.*)

*Teacher*: What were some of the things you noticed that you thought were good about our first attempt at this dance?

*Pupil*: We all seemed to go in the right direction and didn't get mixed up.

*Pupil*: We kept time pretty well.

*Teacher*: What are some of the things you liked about it?

*Pupil*: I like always being in a different circle.
*Pupil*: Well, it made me catch on to fractions better.
*Pupil*: Me too. We ought to do arithmetic that way all the time.
*Teacher*: You think it was easier to learn fractions?
*Pupil*: I'll say, and it was lots more fun.
*Teacher*: What are some of the ways we might improve it if we tried it another time?
*Pupil*: We ought to try to all pop at the same time.
*Pupil*: Maybe we shouldn't try to go around the circle so fast.
*Teacher*: Yes, those are good suggestions. Personally, I think it was rather well done for our first attempt.

The next situation indicates how the learning of addition combinations might be accomplished with a group of children through a game called *Exchange Numbers*. This is a suitable activity for a limited area of space, such as the regular classroom. Take twenty 4″ x 6″ index cards and write a number from 1 to 20 on one side of each card. On the other side, write combinations or addend pairs for the sum on the first side. If the number is 12 on one side, the following combinations could be on the other side:

$$\begin{array}{r} 6 \\ +6 \\ \hline \end{array} \quad \begin{array}{r} 7 \\ +5 \\ \hline \end{array} \quad \begin{array}{r} 8 \\ +4 \\ \hline \end{array} \quad \begin{array}{r} 9 \\ +3 \\ \hline \end{array} \quad \begin{array}{r} 10 \\ +2 \\ \hline \end{array} \quad \begin{array}{r} 11 \\ +1 \\ \hline \end{array} \quad \begin{array}{r} 12 \\ +0 \\ \hline \end{array}$$

Duplicate the twenty cards. (The cards can be made by the teacher or pupils, or both.) Each pupil is given a card. He looks at the number on the card and then turns it over. With the duplicate cards, two pupils will have the same card. The teacher, or a pupil acting as leader, stands in the front of the room and calls out two numbers (a combination). The persons having the sum of these two numbers stand. They attempt to run and exchange seats before being tagged by the leader.

*Teacher*: We have been learning addition facts and today I want to see how well you remember some of them. We are going to play a game called *Exchange Numbers*. In order to play this game, we are going to have to be quick thinkers and quick runners so as not to be caught. What does the name of the game mean to you, James?
*Pupil*: Does it mean to trade with someone?
*Teacher*: Yes, it means trading with someone. Another name

might be *Exchange Seats* because we trade seats with someone else. I am going to give each of you a card. Look at your card on one side and then turn it over. No one else should see your number. What do you see on the other side of your card, Mary?

*Pupil*: Numbers that add up to the number on the other side.

*Teacher*: Yes. Now two children in the room have the same number, so someone else in the room has the same number as you have on your card. We need one person to stand in the front of the room. John, will you be *It*? John will call out two numbers. As soon as he has called out the two numbers, add them together in your head. If the answer is the number on your card, you will stand. The other person with that number will stand also. You will then run quickly and try to exchange seats. You may run in any direction as long as you do not run between seats. As soon as you leave your seats, John will run and try to tag one of you. You must exchange seats before he tags you, because the one who is tagged becomes *It*. Are there any questions?

*Pupil*: Can we look at our card when John calls the numbers?

*Teacher*: Yes, you may. The card is your helper. However, you will probably get the answer more quickly if you just try to remember the sum for the two numbers. Let's try it once before actually playing it. Suppose John calls the numbers 7 and 5. What is the answer, Dick?

*Pupil*: 12.

*Teacher*: Yes, 12. Jane and Sue, you both have the number 12. Run quickly and sit in each other's chairs. John will run and try to tag you. That's right. You were safe, Sue, but Jane was tagged and so she becomes *It* and John takes Jane's card.

(*The pupils participate in the activity and after a while the teacher evaluates it with them.*)

*Teacher*: Well, everyone seemed to have a good time. What did you learn today?

*Pupil*: We learned a new indoor game called *Exchange Numbers*.

*Teacher*: What do you have to know to be a good player?
*Pupil*: We need to know our addition facts.
*Pupil*: We have to think of the answer quickly.
*Pupil*: We must decide the best way to run to the other seat.
*Pupil*: We must run quickly and dodge the person who is *It*.
*Teacher*: Did it make us think?
*Pupil*: I'll say! It made us think quickly.
*Teacher*: Did it help our bodies in any way?
*Pupil*: Running helps our bodies because it is good exercise.
*Pupil*: I liked everything about this game. It's fun.
*Pupil*: Can we play it again soon? We'll learn our addition facts better for the next time.
*Teacher*: Maybe we could play it when we start learning about multiplication.
*Pupil*: I liked running up and down and dodging *It*.
*Teacher*: How could we improve the game? How could we change it to make it better?
*Pupil*: Maybe three or four of us could have the same number rather than just two, then more of us would have a chance to run. Four people would exchange seats and *It* could tag any of the four people.
*Teacher*: That sounds like a good idea. Suppose we make four copies of each card. We could use numbers 1 through 10 or 10 through 20.
*Pupil*: At one time during the game I was up first and had no place to run and exchange seats, so I was tagged before the other person was out of the seat.
*Teacher*: How could we improve this? Does anyone have a suggestion?
*Pupil*: The runner could stand and touch the back of his chair and be safe until the other person stood up, and *It* would have to stay at the front of the room until someone left his seat.
*Teacher*: That sounds like a good idea.
*Pupil*: I think a penalty should be given to the person who does not exchange seats and stays in his place to keep from being tagged.
*Teacher*: I think all of the suggestions are very good. Next time

we play the game we will try to improve it as much as we can and include the suggestions that you made today.

Another procedure shows how multiplying by tens might be used in a game called *Catch the Cane.* The players stand in a circle, with one person designated as *It* in the center of the circle. All of the players in the circle are numbered consecutively. The person in the center holds the cane (wand, yardstick, ball bat, or the like) with his finger tips. The other end of the cane rests on the surface area. He calls a number and releases the cane. The person having that number must run in and grab the cane before it falls to the surface area. If the person whose number is called fails to get the cane before it strikes the surface area, he goes back in the circle. If he catches it, he changes places with the person in the circle. Some form of scoring can be used if so desired.

*Teacher*: Boys and girls, how many of you ever knocked over something by accident and then reached out to catch it before it fell down?

*Pupil*: I bumped into a lamp this morning, but I caught it before it fell to the floor.

*Teacher*: You were quick enough to catch the lamp before it fell all the way to the floor. Well, this morning we are going to play a game called *Catch the Cane,* where we have to catch something before it falls to the floor. Let us form a circle and count off, starting with number 1. I will also have a number and stand in the center of the circle with my finger holding the tip of the upright cane in this manner. When I call your number, run in and try to catch the cane before it falls to the floor, and I will try to run to your place in the circle. Let's try it once. Number 10. Good. George caught the cane, so he comes to the center to release it this time. If he did not catch the cane, I would still be in the center and he would have to go back to the circle. Now George will call a number. Let's see how quickly you can get in and catch the cane if your number is called.

*(The game continues for a time in this manner and the teacher stops the activity to discuss it with the children.)*

*Teacher*: As you know, we are learning how to multiply by 10 and 100. I think that many of you can do it without paper and pencil. I think you can show me by playing the game we just learned. Let's count off again. But this time let's count by 10's instead of starting with 1. (*Pupils count off by 10's*). Now when the person in the center calls out a number, he will call out 10 times a number: 10 times 7, 10 times 8, and so forth.

(*The game continues in this manner and then the teacher evaluates it with the class.*)

*Teacher*: Can you think of any ways in which we could improve this game?

*Pupil*: Well, we could run in for the cane faster.

*Pupil*: We ought to make the circle bigger so we could have farther to run for the cane.

*Teacher*: How about the way we played the game the second time?

*Pupil*: You have to listen more carefully for the numbers because you have to multiply them.

*Pupil*: You have to figure the answer out fast to catch the cane before it falls to the floor.

*Pupil*: I knew the right answer once, but I just didn't get in fast enough. Maybe we should give the answer when we start to run in for the cane.

*Teacher*: You have made some good suggestions. Now can you think of any other ways you might play the game the next time?

*Pupil*: We could do it by *dividing* numbers by 10.

*Teacher*: Yes, that's a fine idea, and we might try it another time.

The final procedure presented here illustrates how a better understanding of fractions might be developed by participation in the game *Triple Change*. All of the players except three form a circle. The remaining three stand in the center of the circle. The players in the circle count off by 3's and the players in the center are numbered 1, 2, and 3. The players in the center take turns calling out their own numbers. When one of them calls a number (1, 2, or 3), the players in the circle who have that number attempt to change places. The player in the center of the circle tries to catch one of the players as they change places. If a player is caught, he changes places with the caller.

*Teacher*: Today we are going to play a game called *Triple Change*. What do you think the word "triple" means? When we think of "triplets" what number do we think of?

*Pupil*: Three?

*Teacher*: Yes, Frank, how did you know that?

*Pupil*: I saw some triplets on television. There were three people who looked alike.

*Teacher*: Now I would like to have three of you be a triplet. (*three children volunteer*) All right. Will you three stand here by me? Do you remember another word that we used in arithmetic class yesterday that is related to three?

*Pupil*: Was it third?

*Teacher*: Yes. Who can write one-third on the board? (*Pupil writes it on the board*) Now how many thirds are there in a whole?

*Pupil*: There are three.

*Teacher*: Three-thirds is right. Three-thirds and one are the same number. (*Teacher writes 3/3 = 1 on the board.*) Now let's play the game. Form a circle. The set of triplets remain here by me. Those of you in the circle count off in groups of three, and remember your number. All right, fine. Now let's have all the 1's raise their hands, the 2's, the 3's. Very good. The triplet is numbered 1, 2, and 3. These children will be callers and they will take turns in calling their numbers. When a number is called, all who have that number must quickly change places with each other. You may do this by changing with a neighbor or running through the circle. The center players take turns calling. If the first one does not get a place, then the second one calls. Should the first succeed in catching one, the player caught will wait his turn in the center until number 2 and number 3 have had a turn at calling before he calls a number. Let's try it once for practice. Fred, call your number.

(*After the demonstration, the game is played for a time, and then the teacher evaluates it with the class.)*

*Teacher:* What were some of our problems?

*Pupil:* Some players did not remember their numbers and ran at the wrong time.

*Pupil:* Sometimes it was hard to find a place.

*Pupil:* Some kids didn't run when their number was called.

*Teacher:* Well, now, what do you suppose we could do about some of those things?

*(After some discussion, the following decisions were made by the group.)*

1. The players in the circle could hold hands, letting only the numbers called go.
2. The players who did not move, or who moved at the wrong time, would have to pay a forfeit.
3. Also, when caught three times, a forfeit would be required. The purpose of this was to discourage players from deliberately being caught in order to become a caller.

The illustrations that have been presented here should give some idea of how learning of mathematics can take place in actual teaching-learning situations. Perhaps these illustrations can serve as a guide for teachers in planning their own lessons.

Chapter 4

# RESEARCH IN LEARNING ABOUT MATHEMATICS THROUGH MOTOR LEARNING

IT was mentioned in Chapter 2 that learning through motor activity is not necessarily a recent innovation. In fact, over the years the literature on this general area has been replete with pronouncements of eminent philosophers and educators. A representative sampling of these comments follows:

*Plato* (380 B.C.) Lessons have been invented for the merest infants to learn, by way of play and fun . . . Moreover, by way of play, the teachers mix together (objects) adapting the rules of elementary arithmetic to play. (This is particularly applicable to the topic hand).

*Aristotle* (350 B.C.) It should not be forgotten that it is through play that the path is opened toward occupations of later age, and it is for this reason that the majority of games are imitations of work and actions which will be used in later life.

*Quintillian* (A.D. 100) Play . . . is a sign of vivacity, and I cannot expect that he who is always dull and spiritless will be of an eager disposition in his studies, when he is indifferent even to that excitement which is natural to his age.

*Comenius* (1650) Intellectual progress is conditioned at every step by bodily vigor. To attain the best results, physical exercise must accompany and condition mental training.

*Rousseau* (1750) If you would cultivate the intelligence of your pupil, cultivate the power that it is to govern. Give his body continual exercise; make him robust and sound in order to make him wise and reasonable.

*Friedrich Froebel* (1830) It is by no means, however, only the physical power that is fed and strengthened in these games; intellectual and moral power, too, is definitely and steadily gained and brought under control.

*Herbert Spencer* (1860) We do not yet sufficiently realize the truth that as, in this life of ours, the physical underlies the mental.

*G. Stanley Hall* (1902) For the young, motor education is cardinal, and is now coming in due recognition, and for all, education is incomplete without a motor side, for muscle culture develops brain centers as nothing else yet demonstrably does.

*John Dewey* (1919) Experience has shown that when children have a chance at physical activities which bring their natural impulses into play, going to school is a joy, management is less of a burden, and learning is easier.

*L. P. Jacks* (1932) The discovery of the educational possibilities of the play side of life may be counted one of the greatest discoveries of the present day.

Thus spoke some of the most profound thinkers in history. The favorable pronouncements of such people with regard to the possibilities of intellectual development through motor activity obviously carry a great deal of weight. However, in an age when so much emphasis is placed upon scientific inquiry and research, we cannot accept only the subjective opinions of even some of history's most profound thinkers; thus, the necessity to place an objective base under a long-held theoretical postulation is established.

## RESEARCH TECHNIQUES

How does one go about studying the motor activity learning phenomenon to see if learning actually does take place through this medium, and how well it compares with some of the traditional ways of learning?

There are a number of different ways of studying how behavioral changes take place in children. After considerable study and experimentation, a certain sequence of techniques emerged as the most appropriate way to evaluate how well children might learn through motor activity. These techniques are generally identified as follows:

1. naturalistic observation

2. single-group experimental procedure
3. parallel-group experimental procedure
4. variations of standard experimental procedures

## Naturalistic Observation

In our work in this area, one of the first problems to be reckoned with was whether this type of learning activity could be accomplished in the regular school situation, and also whether teachers whose preparation and experience had been predominantly in traditional methods would subscribe to this particular approach. To obtain this information, a procedure that could best be described as "naturalistic observation" was used. This involved the teaching of a skill or concept in mathematics to a group of children using a motor activity in which the skill or concept was inherent. The teacher would then evaluate how well the skill or concept was learned through the motor activity learning medium. The teacher's criteria for evaluation were his or her past experiences with other groups of children and other learning media. An example of a case of naturalistic observation follows:

*Concept*: Dividing to find the number of sets.

*Activity*: Get Together. Players take places around the activity area in a scattered formation. The leader calls any number by which the total number of players is not exactly divisible. All players try to form groups of the number called. Each group joins hands in a circle. The one or ones left out can have points scored against them if a scoring procedure is desired. Low score wins the game after it is played for a specified amount of time. As an example, if there were 23 children and the teacher called *five,* there would be 20 of the children who would hurry to make up four groups with three children being left over.

*Teacher Evaluation*: The activity is useful for reinforcing the idea of grouping. It gave the children the idea that there may be a *remainder* when dividing into groups.

Naturally, this procedure is grossly lacking in objectivity because there is only a subjective evaluation of the teacher to support the hypothesis. However, in the early stages of the work, this technique served our purpose well because at that time we were mainly concerned with having teachers experiment with the idea and ascertaining their reactions to it. In a vast majority of cases the reactions of teachers were very positive.

## Single Groups

The next factor that needed to be taken into consideration was whether or not children could actually learn through the motor activity learning medium. Although for centuries empirical evidence had placed the hypothesis in a very positive position, there was still the need for some objective evidence to support the hypothesis. In order to attempt to determine if learning could actually take place through motor activity, the *single group technique* was employed. This technique involved the criterion measure of objective pretesting of a group of children on certain skills or concepts in mathematics. Motor activities in which mathematics skills or concepts were inherent were taught to the children over a specified period of time and used as learning activities to develop the skills or concepts. After a specified period of time the children were retested and the results of the posttest were compared with the results of the pretest.

All of our studies involving this technique in which the subjects were their own controls have shown significant differences between pretest and posttest scores at a very high level of probability. Therefore, it appeared reasonable to generalize that learning actually could take place through the motor activity learning medium.

## Parallel Groups

With the preceding information at hand, the next and obviously most important step in the sequence of research techniques was to attempt to determine how the motor activity learning medium compared with other more traditional learning media. For this purpose the *parallel group technique* was used. This involved pretesting children on a number of skills or concepts in mathematics and dividing them into two groups on the basis of pretest scores. One group would be designated as the motor activity group (experimental group) and an attempt would be made to develop the skills or concepts through the motor activity learning medium. The other group would be designated as the traditional group (control group), with an attempt made to develop the skills or concepts through one or more traditional media. Both groups would be taught by the same classroom teacher over a specified period of time. At the end of the experiment both groups would be retested and comparisons made of the posttest scores of both groups.

## Variations of Standard Experimental Procedures

A number of additional variations of standard techniques have also been employed. For example, in studying the effectiveness of the motor activity learning medium for boys compared to girls, a procedure was used that involved parallel groups of boys and girls within the total single group.

In those cases where an attempt has been made to hold a certain specific variable constant, three groups have been used. In this situation one group becomes an observing or nonparticipating group.

Another variation has been to divide children into two groups with each group taught by a different teacher. This can be done for the purpose of comparing the physical education teacher, who would not likely be skilled in teaching concepts in another curriculum area, with a superior classroom teacher who would likely be highly skilled in this direction.

In the authors' studies, the experiment is usually carried on over a period of ten days. In some cases where conditions would permit, this time period has been longer. There are ordinarily eight and sometimes as many as ten skills or concepts involved. A ten-day period allows for two days of testing and eight days of teaching. Reliability for the objective tests has ordinarily been obtained by using a test-retest with similar groups of children. All of the authors' experiments have taken place in the actual school situation. Obviously, it would be better to carry them out over extended periods of time, but in most cases it has been impractical to do so because it usually involves some interruption in the regular school program. In addition, it should be mentioned that our studies are much more exploratory than they are definitive; this is ordinarily the case when conducting many kinds of research involving young children.

## SOME REPRESENTATIVE RESEARCH FINDINGS

The first study reported here is an example of a single group experimental procedure with parallel groups of boys and girls within the single group.[1] The purpose of this study was to determine how well a group of first grade children might develop number concepts through motor activity in the form of active games, and at the same time to ascertain whether the approach was more favorable for boys or for girls.

Thirty-five first grade children were pretested on eight number concepts which were to be included as a part of their regular classwork during the ensuing two weeks. Ten boys and ten girls who had the same pretest scores as the boys were selected for the experiment. Eight active games in which the number concepts were involved and which were appropriate for use at first grade level were selected.

The active games were taught to the twenty children and used as learning media for the development of the number concepts. They were retested after the active game medium was used. The scores of the first test for all of the twenty children ranged from 30 to 73 and the scores on the second test from 59 to 78. The mean score of the first test was 51.7 and the mean score of the second test was 68.0. In computing the results for the ten boys and the ten girls separately, the mean score for the second test for the boys was 70.5, and for the girls 65.5. The statistical analysis showed that as a total single group, there was a highly significant difference between pretest and posttest mean scores. In comparing boys with girls, the results indicated greater changes in learning were produced with the boys.

---

[1]Humphrey, James H., "An exploratory study of active games in learning of number concepts by first grade boys and girls," *Perceptual and Motor Skills, 23*: 1966.

For a clearer picture of the direction of differences between the boys and girls, the following table shows the percentages of differences in gain on a paired per pupil basis:

| *Subject Pair* | *Differences in % of Gain* | *Sex* |
|---|---|---|
| 1 | 0 | - |
| 2 | 11 | Boy |
| 3 | 0 | - |
| 4 | 8 | Girl |
| 5 | 41 | Boy |
| 6 | 40 | Boy |
| 7 | 31 | Boy |
| 8 | 34 | Boy |
| 9 | 64 | Boy |
| 10 | 28 | Girl |

Six of the boys had a greater percentage difference in gain than the girls, while two of the girls had a greater percentage difference in gain. In two cases there was no difference.

In the next study reported here, a large number of pupils were involved in comparing motor activity learning by active games with two other procedures in developing concepts related to the telling of time.[2]

Forty-two classes of third grade pupils, a total of 1,147 children from eighteen school districts, were used as subjects in this study. The original parent population from which the forty-two classes were randomly selected consisted of 319 third grade classes from 166 elementary schools.

The forty-two classes were divided into three groups of fourteen each. One group was taught through the developmental-meaningful method. A second group was taught through the drill method, and with the third group the active game approach was used. All classes were taught by their own classroom teacher.

Three sets of lesson plans, one for each group, were devised.

---

[2]Crist, Thomas, *A comparison of the use of the active game learning medium with developmental-meaningful and drill procedures in developing concepts for telling time at third grade level,* Doctoral Dissertation, University of Maryland, College Park, Maryland, 1968.

Instructions to the teachers were included in each set of lesson plans. Lesson plans for the developmental-meaningful group closely followed the objectives, suggested activities, and problems used in the several textbooks most prevalent in the area. Lesson plans for the drill group and active game group paralleled the materials covered in the developmental-meaningful group. In the lesson plans for the drill group, the suggestions took the form of having the pupils work on prepared examples in individual drill booklets. In the lesson plans for the active game group, the suggestions took the form of pleasurable active games. A twelve-foot clock was painted on the playground of each school using the active game procedure, and materials needed for the implementation of the program were made available. Included in this equipment were four playground balls, two sets of flash cards, and one set of numbered blocks.

There were ten teaching days in the experiment. All teachers were required to teach each lesson in twenty-minute periods. A lesson began immediately after the teacher had read the stated objectives for the current lesson. All teachers taught the time-telling lesson at the same time each day. Teachers who used the active game approach considered time spent on lesson plans for this experiment as part of their arithmetic class rather than part of their physical education period.

Two parallel forms (Forms A and B) of a performance test in time-telling concepts were constructed and used as criterion measures for the study. Each form of the criterion measure contained seventy-four items divided into two main parts. Part I of each form consisted of sixty items and purported to measure primarily a basic understanding of time telling and comprehension of the passing time. Part II of each form consisted of fourteen verbal problems. Form A was administered as a pretest and Form B was administered as a posttest.

In comparing the pretest and posttest of each individual group as its own control, it was found that all groups learned from pretest to posttest. However, the highest level of probability was observed in the active game group, the second highest in the developmental-meaningful group, followed by the drill group. When a comparison was made of the posttest scores of all three groups, there was no significant difference between any of the three groups.

In view of the fact that none of the teachers taking part in the experiment had ever taught an academic concept through the use of motor activity, it seemed reasonable to consider that any conclusions drawn must necessarily take this factor into consideration. It was also necessary to assume that all of the teachers had some, if not considerable, preparation and experience in the use of the developmental-meaningful and drill teaching procedures. Therefore, it would appear justifiable to speculate what results would have been obtained if a similar experiment could be carried out using teachers with a motor activity-oriented teacher preparation background to teach time telling on the playground instead of teachers only with a general elementary education background. And further, because paper and pencil tests were used, the whole experimental testing procedure could be viewed as slanted toward the two traditional classroom procedures. Again, one could speculate as to the results if the testing (posttest) had been administered under one of the following circumstances: (1) testing the pupils in all three procedures in an active game locale, (2) testing the pupils taught by the active game procedure with a paper and pencil test and testing pupils taught by the developmental-meaningful and drill procedures in an active game situation, and (3) testing pupils taught by the active game procedure in an active game locale and testing pupils taught by the other two procedures with a paper and pencil test.

The purpose of the next study reported here was to attempt to determine the effect of active games in comparison with passive games as an approach to providing learning experiences designed to develop arithmetic readiness skills and concepts at the kindergarten level.[3] Furthermore, the intent was to provide information as to whether these activities at the kindergarten level would have an effect on the development of arithmetic readiness skills and concepts in comparison with the traditional teaching procedures peculiar to kindergarten.

---

[3]Droter, Robert, *A comparison of active games and passive games used as learning media for the development of arithmetic readiness skills and concepts with kindergarten children in an attempt to study gross motor activity as a learning facilitator,* Master's thesis, University of Maryland, College Park, Maryland, 1972.

Sixty children were randomly placed in three groups of twenty, with one group taught through active games, one group taught through passive games, and one group taught through traditional procedures. On the basis of comparison of pretest and posttest scores, the findings indicated that (1) learning took place in all groups at a high level of probability, and (2) the active game group had the highest mean gain although not significantly better than the other groups. On the basis of the statistical findings and subjective observations of the teachers, the investigator made the following generalizations: It was evident that kindergarten children learned arithmetic readiness skills and concepts from the normal instructional approaches peculiar to kindergarten. However, the addition of games, both active and passive, would improve the arithmetic readiness development of kindergarten children. Specifically, the findings showed that active games facilitated learning as well as or even better than the other approaches. Thus, it would seem that teachers could use game and play activities in such a way as to provide participation in arithmetic readiness skills and concepts with kindergarten children. (It is interesting to note that the investigator, himself an elementary school physical education teacher, recommended that elementary school physical educators might be professionally prepared in the use of motor learning activities so that they could possibly function as resource personnel in the development of mathematics skills and concepts.)

In the next study sixty children, five- and six-year-olds, were used as subjects to compare certain techniques in teaching mathematics.[4] They were divided into three equal groups on the basis of an eighty-item verbal test on eight mathematical concepts.

One group received all of its number instruction in the traditional classroom manner. The second group was taught by way of motor learning activities. The formation of the third group was an attempt to investigate the effect of isolating the motivation aspect of motor activity learning. This group was permitted to watch and listen but not participate actively in the instruction

---

[4]Wright, Charles, *A comparison of the use of traditional and motor activity learning media in the development of mathematical concepts with five and six year old children, with an attempt to negate the motivational variable,* Master's thesis, University of Maryland, College Park, Maryland, 1969.

given to the motor activity group.

The groups were taught one concept per day for eight consecutive school days and then given the posttest. There were no statistically significant differences between any of the groups. However, when each group was used as its own control it was observed that the motor activity group learned at a very high level of probability. The investigator felt that this was an impressive gain due to the fact that the classroom teacher had no previous experience in using the motor activity learning medium.

Another study attempted to determine whether or not motor learning activities could be used as a practical and effective enrichment aid in the teaching of selected first grade mathematical concepts.[5]

Two units of the Book I program of a standard mathematics series were reviewed and ten concepts were selected. From these concepts, an objective test containing five items per concept was developed. Each item was rated for difficulty and questions which received a score below .30 and above .75 on the difficulty rating were reworded.

The test was administered to the subject class and using the results of this pretest, the subject class was divided into two equal groups. Motor activity learning experiences were then selected to enrich each of the mathematical concepts.

The *entire* class received traditional classroom teaching procedures in learning the mathematical concepts involved. After each lesson, the enrichment group (experimental group) was given enrichment by the physical education teacher through the use of a variety of physical education activities. The control group participated in free play supervised by the classroom teacher during this time.

After four weeks of the above procedure, a posttest was administered to both groups. After an additional four weeks without enrichment, an extended interval test was administered to both groups.

In comparing the pretest scores, no significant differences were

---

[5]Krug, Frank, *The use of physical education activities in the enrichment of learning of selected first-grade mathematical concepts*, Master's thesis, University of Maryland, College Park, Maryland, 1973.

found between the groups as a whole and for each sex separately. Significant differences were found at a very high level of probability between the scores of the posttest for the groups as a whole and for each sex separately, with the enrichment group learning more than the control group. When comparing the extended interval tests, no significant differences were found between groups for girls and boys separately; however, when considered as an entire group, significance was found at a high level of probability favoring the enrichment group.

Further interpretations of the data indicated that the results of the pretest showed that both control and enrichment groups started out statistically equal. This was observed for all groups (girls only, boys only, and boys and girls combined.) The results of the posttest showed that all control and enrichment groups gained in the mathematical concepts at a very high level of probability. The results of the posttest further showed that all enrichment groups learned significantly more than the control groups. The results of the extended interval tests showed that all groups retained what they had learned. The enrichment group more effectively retained their learning over the control group when examined as boys and girls combined.

In addition to the statistical analysis, the following subjective evaluation was made by the classroom teacher:

> The children had difficulty understanding why only half of them went with their physical education teacher after mathematics. By this, I mean the enrichment group felt privileged while the control group felt left out. However, they came to accept this after a few days. Those who were in the enrichment group looked forward to going with the physical education teacher. They enjoyed the games immensely. Now that the experiment is completed, the children in the enrichment group have shown me many of the games they played with the physical education teacher and how they related to the concepts being taught in class. We are now using some of the games as reinforcement for other concepts. I have found this extremely beneficial. In my estimation, the experiment was well planned, well conducted, and of value to the children.

Generalized conclusions drawn from this study were that the use of motor learning activities was effective in enriching the mathematics concepts involved and should be considered as an enrichment aid when teaching first grade mathematical concepts. The classroom teacher recommended that physical education teachers might well be considered as important consultants in the planning of certain types of learning experiences relevant to first grade mathematical concepts. In other words, the physical education teacher could be considered as a valuable coworker with the classroom teacher in the development of mathematical concepts.

The purpose of the final study reported here was to explore statistically the relationships between the traditional and physical activity techniques the classroom teacher and physical education teacher use in teaching selected mathematical concepts to first grade children.[6] A secondary purpose was to see if first grade children could learn selected mathematical concepts through the physical activity technique when taught by the physical education teacher.

A fifty-two item test dealing with number concepts from 2 to 9 was constructed by the school principal, first grade teachers, and the investigator. The test was constructed to measure the number knowledge of a group of first grade children who scored low on the numbers section on a standardized readiness test administered during the second week of school. The results of this numbers readiness test were used in grouping first grade pupils for arithmetic. The low group, which was used for the study, was primarily developed from the first grade pupils scoring eight and below on the numbers section. This group consisting of forty pupils was tested and equated on the basis of the results of the test.

The regular first grade classroom teacher taught the number concepts to the control group by traditional classroom techniques and the experimental group was taught by the physical education teacher by use of motor learning activities. The teaching time for the two groups involved nine class periods of thirty minutes each. At the conclusion of the nine class periods the two groups were retested to see if there were statistically significant differences in the two teaching techniques as taught by the classroom teacher

---

[6]Trout, Edwin, *A comparative study of selected mathematical concepts developed through physical education activities taught by the physical education teacher and traditional techniques taught by the classroom teacher*, Master's Thesis, University of Maryland, College Park, Maryland, 1969.

and the physical education teacher.

In the experimental group taught by the physical education teacher, the mean score of the first test was 36.55, and for the second test 44.44, for a raw mean increase of 7.89. A statistical analysis of these data indicated a difference at an extremely high level of probability.

The control group taught by the classroom teacher had a mean score of 38.16 for the first test and a mean score of 43.52 for the second test, for a gain of 5.36. There was also a significant difference for this group but at a lower level of probability than for the experimental group. When the posttest scores of both groups were compared there was a significant difference in favor of the experimental group at a moderately high level of probability.

Any conclusions to be drawn from a study of this nature must be governed by a degree of caution. However, the following generalizations appeared warranted by the investigator:

1. The experimental group indicated a trend toward a more uniform rate of learning as indicated by the standard deviations of both groups.
2. Physical education activities could be used with success in helping slower learning first grade children learn about mathematics.
3. The physical education teacher could serve as a consultant when formulating the first grade arithmetic program.

## Some Generalizations of the Research Findings

It should be mentioned again that the research in this general area is much more exploratory than definitive. However, it is interesting to note that no study has shown a significant difference in favor of traditional procedures over the motor activity learning procedure.

In view of the fact that there are now some objective data to support a long-held hypothetical postulation, perhaps some generalized assumptions along with some reasonable speculations can be set forth with some degree of confidence. Obviously, the available data reported in the foregoing studies are not extensive enough to carve out a clear cut profile with regard to learning through motor activity. However, they are suggestive enough to give rise to some interesting generalizations, which may be briefly summarized as follows:

1. In general, some children tend to learn certain mathematics skills and concepts better through the motor activity learning medium than through many of the traditional media.
2. This approach, while favorable for both boys and girls, appears to be more favorable for boys.
3. The approach appears to be more favorable for children with average and below-average intelligence.
4. Many teachers report that for children with high levels of intelligence, it may be possible to introduce more advanced skills and concepts at an earlier age through the motor activity learning medium.

It will be the responsibility of research to provide the conclusive evidence to support these generalizations and speculations. There is hope, however, based on actual experience with this approach in the activities described throughout this text, and particularly in the final three chapters, to encourage those responsible for facilitating the learning of mathematics skills and concepts by children to use this approach and to join in collecting evidence to verify the contribution of motor activity learning to the education curriculum.

Chapter 5

# MATHEMATICS MOTOR ACTIVITY STORIES

## MOTOR-ORIENTED READING CONTENT

BASIC facts about the nature of human beings serve educators today as principles of learning. One of these principles — that the child's own purposeful goals should guide his learning activities — serves as the basis for developing motor-oriented reading content material.

One of the earliest and perhaps the first attempt to prepare motor-oriented reading content as conceived here is the work of Humphrey and Moore.[1] This original work involved a detailed study of reactions of six- to eight-year-old children when independent reading material is oriented to active game participation. The experiment was initiated on the premise of relating reading content for children to their natural urge to play.

Ten games were written with a story setting that described how to play the games. The manuscripts were very carefully prepared. Care was given to the reading values and the literary merits of each story. Attention was focused upon (1) particular reading skills, (2) concept development, (3) vocabulary load, that is, in terms of the number, repetition and difficulty of words, and (4) length, phrasing, and number of sentences per story.

When the manuscripts were prepared, the *New Readability Formula for Grades I-III* by George D. Spache was applied to judge the reading difficulty of the material. Following this, thirty teachers in rural, suburban, and city school systems working with fifty-four reading groups of children used and evaluated the stories in actual classroom situations. The children represented to a reasonable extent a cross section of an average population with

[1]Humphrey, James H., and Moore, Virginia D., "Improving reading through physical education," *Education* (The Reading Issue) May, 1960.

respect to ethnic background, socioeconomic level, and the like. In all, 503 children read from one to three stories for a total of 1,007 different readings.

On report sheets especially designed for the purpose, the teachers were asked to record observable evidence of certain comprehension skills being practiced by the reading groups before, during, and after the children played the games they read about. The teachers were requested to make their evaluations on a comparative basis with other materials that had been read by the children. The results of these observations were as follows:

| *Comprehension Skills* | *Per Cent of Groups Practicing Skills* |
|---|---|
| Following Directions | 91 |
| Noting and Using Sequence of Ideas | 76 |
| Selecting Main Idea | 76 |
| Getting Facts | 67 |
| Organizing Ideas | 46 |
| Building Meaningful Vocabulary | 41 |
| Gaining Independence in Word Mastery | 35 |

In another dimension of the study teachers were asked to rate the degree of *interest* of the children in the reading on an arbitrary 5-point scale as follows: extreme interest, considerable interest, moderate interest, some interest, or little or no interest. The fact that there was sustained interest in the active game stories is shown in the following results. (The 25 cases in the last two categories involved children with IQ's far below normal.)

| *Degree of Interest* | *Number and Percent of Cases* |
|---|---|
| Extreme Interest | 469-46% |
| Considerable Interest | 242-24% |
| Moderate Interest | 271-27% |
| Some Interest | 22-2.7% |
| Little or no interest | 3- .3% |

The preceding results become more meaningful when it is considered that many of the classroom teachers reported that untold numbers of children sit in school and read with little or no interest. This dimension of the study tended to verify that reading is an active rather than a passive process. Apparently the children had a real and genuine purpose for reading. To satisfy their natural urge to play they became interested and read to learn how to play a new game.

On the basis of the findings and the limitations involved in conducting such an experiment, the following tentative conclusions seemed warranted:

1. When a child is self-motivated and interested, he reads. In this case the reading was done without motivating devices such as picture clues and illustrations.
2. These game stories were found to be extremely successful in stimulating interest in reading and at the same time improving the child's ability to read.
3. Because the material for these stories was scientifically selected, prepared, and tested it is unique in the field of children's independent reading material. The outcomes were most satisfactory in terms of children's interest in reading content of this nature.

As a result of this study, additional stories were written (131 in all) and developed into a series of six books for first and second grade children.[2]

Widespread success resulting from the use of this material inspired the development of the same general type of reading content which would also include mathematics experiences. This kind of reading content was arbitrarily called *the mathematics motor activity story*.

## EXPERIMENTS WITH MATHEMATICS MOTOR-ORIENTED READING CONTENT

Early attempts to develop mathematics motor activity stories

---

[2]Humphrey, James H. and Moore, Virginia D., *Read and Play* series, London, Frederick Muller, Ltd., 1965. Another version of this work appears as Humphrey, James H., *Learning to Listen and Read through Movement*, Deal, New Jersey, Kimbo Educational, 1974.

were patterned after the original procedure used in providing for motor-oriented reading content. That is, several stories were written around certain kinds of motor activities, the only difference being that the content also involved reference to mathematics experiences. These stories were used in a number of situations. It soon became apparent that with some children the internalization of the mathematics concept(s) in a story was too difficult. The reason for this appeared to be that certain children could not handle both the task of reading while at the same time developing an understanding of the mathematics aspect of the story. It was then decided that since *listening* is a first step in learning to read, auditory input should be utilized. This process involved having children listen to a story, perform the activity, and simultaneously try to develop the mathematics concept. When it appeared desirable, this process was extended by having the children read the story after having engaged in the activity. The following is a description of one of the authors' first experiences with a mathematics motor activity story.[3]

The following story was written, used, and evaluated by a first grade teacher with her class of thirty children. The name of the story is "Find a Friend" and it is an adaptation of a game called *Busy Bee*. The readability level of the story is 1.5 (fifth month of first grade). The mathematics concepts inherent in the story are: *groups or sets of two, counting by twos, and beginning concept of multiplication.*

*Find a Friend*

In this game each child finds a friend.
Stand beside your friend.
You and your friend make a group of two.
One child is *It.*
He does not stand beside a friend.
He calls, "Move!"
All friends find a new friend.
*It* tries to find a friend.
The child who does not find a friend is *It.*

---

[3]Humphrey, James H., "The mathematics motor activity story," *The Arithmetic Teacher*, January 1967.

Play the game.
Count the number of friends standing together.
Count by two.
Say two, four, six,
Count all the groups this way.

The group of first grade children with whom this experiment was conducted bordered on the remedial level and had no previous experience in counting by twos. Before the activity, each child was checked for the ability to count by two's and it was found that none had it. Also, the children had no previous classroom experience with beginning concepts of multiplication.

The story was read to the children and the directions were discussed. The game was demonstrated by the teacher and five pairs of children. As the game was being played, the activity was stopped momentarily and the child who was *It* at that moment was asked to count the groups by two's. The participants were then changed, the number participating changed, and the activity was repeated.

In evaluating the experiment it was found that this was a very successful experience from a learning standpoint. Before the activity none of the children were able to count by two's. A check following the activity showed that eighteen of the thirty children who participated in the game were able to count rationally to ten by two's. Seven children were able to count rationally to six, and two were able to count to four. Three children showed no understanding of the concept. No attempt was made to check beyond ten because in playing the game the players were limited to numbers under ten.

There appeared to be a significant number of children who had profited from this experience in a very short period of time. The teacher maintained that in a more conventional teaching situation the introduction and development of this concept with children at this low level of ability would have taken a great deal more teaching time and the results probably would have been attained at a much slower rate.

Several experiments similar to the one presented above were conducted with much the same results. With such information at hand it was decided to attempt a more objective approach in order to compare the use of the mathematics motor activity story with traditional ways of teaching mathematics. To this end the following study was conducted.[4]

Two groups of second grade children with twenty-one in one group and twenty-three in the other group were pretested on two-number addition facts, three-number addition equations, and subtraction facts. The test contained forty-five items.

One group of children, designated as the experimental group, was taught through six mathematics motor activity stories of the type illustrated previously. The other group, designated as the control group, was taught through such traditional procedures as the printed number line, plastic discs, and abstract algorisms. The experiment was conducted over a four-day period and both groups were taught by the same classroom teacher. Each group was characterized by heterogeneity as far as age and IQ were concerned.

At the end of the four-day period both groups having been taught the processes as indicated above were retested. Comparison of the pre- and posttest scores were evaluated, and at an extended interval of ten days after the posttest, the test was given again with the same statistical procedures applied to the posttest and extended interval test.

The range of scores on the pretest for the experimental group was 12 to 44 with a mean of 30.48, and for the control group 4 to 44 with a mean of 30.87. The posttest scores of the experimental group ranged from 18 to 45 with a mean of 37.77, while the posttest scores of the control group ranged from 5 to 44 with a mean of 34.91. In the experimental group the range of scores on the extended interval test was from 7 to 45 with a mean of 40.50, and in the control group from 15 to 45 with a mean of 39.17. The gain from pretest to posttest favored the experimental group at a high level of probability. Essentially the same results were obtained in gain from posttest to extended interval test.

In recognition of various limitations imposed by a study of this nature any conclusions should be characterized by caution. If one

[4]Humphrey, James H., "Comparison of the use of the physical education learning medium with traditional procedures in the development of certain arithmetical processes with second grade children," *Research Abstracts* A.A.H.P.E.R., Washington, D. C., 1968.

accepts the levels of significant differences in the test scores as evidence of learning, these second grade children could better develop certain computational skills and perhaps retain them longer through use of the mathematics motor activity story than through some of the traditional procedures.

Finally, it seems pertinent to mention certain observations which would not show up in the statistical analysis. It was noticed that the children in the experimental group appeared to be stimulated by the use of the mathematics motor activity story. In fact, some of them commented, "We didn't have arithmetic today." This could possibly mean that the learning activities were enjoyed to the extent that the children might not have been aware of the particular number skills they were using. It is worthy to note also that several of the children in the experimental group seemed to have had little or no interest in any of their arithmetic work until after they got into the experience provided by the mathematics motor activity stories.

Another study (unpublished) employed exactly the same procedures as the one reported above with the exception that it was conducted with slow-learning children and the number of items extended to sixty. The average IQ for the experimental group was 83 and the average IQ for the control group was 86. The results of the comparisons of the posttest scores and extended interval scores favored the experimental group at a very high level of probability.

## MATERIALS RESULTING FROM EXPERIMENTS

After considerable experimentation with the mathematics motor activity stories as indicated in the preceding discussion, a compilation has been made of materials.[5]

It consists of two long-playing records of mathematics motor activity stories and a teacher's manual which provides suggestions for use of the material. The manual provides for general ways to use the material and more detailed specific ways for each individual selection. When a teacher plans to use a given selection, all of the information connected with it should be read very

[5]Humphrey, James H., *Teaching Children Mathematics Through Games Rhythms, and Stunts,* LP No 5000, Deal, New Jersey, Kimbo Educational, 1968.

carefully. This information consists of (1) a story written at a specific reading level, (2) the mathematical skills, concepts, and learnings inherent in the story, and (3) suggested ways of developing the skills, concepts, and learnings with children.

The next step is for the teacher to listen very carefully to the selection before the children hear it, in order to determine the amount of teacher guidance that will be necessary. This may range from little or no teacher guidance to a considerable amount of teacher participation, depending of course upon the particular group of children. The teacher may wish to reproduce the story and have children read it after participating in the activity. It has been found in many cases that certain children can improve their comprehension skills as well as learn the mathematical concepts. Others find the materials valuable in reinforcing certain concepts after they have been introduced. Still others have found it best to use this approach almost entirely, particularly when dealing with slow-learning children. It is highly recommended that teachers should draw upon their own resourcefulness, ingenuity, and imagination as they prepare to use the materials with a particular group of children. The following are some representative examples of the material.

* * * * *

The first example involves the game *Mouse Trap*; the story written about the game has a reading level of 1.7. All of the reading levels shown are based on a readability formula, but they should be considered approximate. It is also possible that the absence of illustrations and picture clues may make a story more difficult for some children to read.

*Mrs. Brown's Mouse Trap*

Some of the children stand in a ring.
They hold hands.
They hold them high.
This will be a mouse trap.
The other children are mice.
They go in and out of the ring.
One child will be Mrs. Brown.
She will say, "Snap!"

Children drop hands.
The mouse trap closes.
Some mice will be caught.
Count them.
Tell how many.
Tell how many were not caught.
Play again.

*Mathematical Skills and Concepts*

Rational counting
More than
Fewer than

*Teaching Suggestions*

1. All of the children can count together the number caught.
2. All of the children can count together the number not caught.
3. If the teacher wishes, he or she can have the children who are caught stand in a line facing those not caught with individual children opposite one another. This way the children can easily see which group has more children. They can also count how many more are in one group than the other.

* * * * *

The next example is concerned with three animal imitation stunts, *Bear Walk, Elephant Walk,* and *Frog Jump.* The reading level is 1.7.

MOVE LIKE ANIMALS

We try to move like animals.
We move like a bear.
We move like an elephant.
We move like a frog.
Try to move like these three animals.
Take 5 steps like a bear.
Take 4 steps like an elephant.
Take 2 jumps like a frog.
Now do it the other way.
Take 2 jumps like a frog.
Take 4 steps like an elephant.

Take 5 steps like a bear.
You did it the other way.
You did the same number of steps and jumps.
Do it in other ways.
Move like other animals.

*Mathematical Skills and Concepts*

Rational counting
Cardinal number ideas
Addition
Commutative and associative laws

*Teaching Suggestions*

1. The children can count the animal movements as they make them. Also they can add to find how many movements altogether.
2. To use the cardinal number idea, after they take five steps like a bear ask, "How many steps did we take? Yes, five."
3. To reinforce the understanding of commutative law the teacher can put 5 + 4 and 4 + 5 on flash cards before and/or after the activity.

## STORIES PREPARED BY TEACHERS

It is highly recommended by the authors that teachers draw upon their own ingenuity and creativity to prepare mathematics motor-oriented stories. One of the reasons for this is the great lack of published materials in this general area. In addition, teachers themselves can provide materials to meet the specific needs and interests of children in a given situation. Of course, a serious drawback is that preparation of such materials is such a time-consuming effort that it becomes more expedient to use professionally prepared materials. However, it has been the authors' experience that those teachers who have the ability and have been willing to take the time have produced amazingly creative stories using games, stunts, and rhythms. In writing such stories using a motor activity setting there are several guidelines that the teacher should keep in mind.

In general the new word load should be kept relatively low, and there should be as much repetition of these words as possible and appropriate. Sentence length and lack of complexity in sentences should be considered in keeping the level of difficulty of material within the independent reading levels of children. This is all the more important if the teacher decides not to use auditory input before the selection is put into the hands of the children to try to read. There are numerous readability formulas that can be utilized. For primary level stories *Spache's Readability Formula*[6] and *MaGinnis' revision of Fry's Readability Graph*[7] are well suited. For upper-level stories, Fry's Readability Graph is useful.

Consideration must also be given to the reading values and literary merits of the story; using a character or characters in a story setting helps to develop interest. The activity to be used in the story should *not* be readily identifiable. When the children identify an activity early in the story there can be resulting minimum attention on the part of the children to get necessary details in order to perform the activity. This also may cause attention distraction from the mathematics concepts inherent in the story. In developing a story, therefore, it is important that the nature of the activity and procedures of the activity unfold gradually.

In developing an active game story the equipment, playing area, and procedures should be clearly described if necessary. Physical education terminology can be used in describing the game setting. In a file, row, or column the children stand one behind the other; in a line the children stand beside each other. Basic motor skills that can be utilized in stunt and rhythmic activities as well as for games include (1) *locomotor skills* (walking, running, leaping, jumping, hopping, skipping, galloping, and sliding); (2) *skills of propulsion* (throwing and striking with underarm, sidearm, and overarm swing patterns); (3) *skills of retrieval* (catching); and (4) *axial skills* (such as twisting, turning, and stetching).

Games should be at the developmental level of children. At the primary level, games should involve a few simple rules and in

---

[6]Spache, George D., *Good Books for Poor Readers,* (Champaign, Illinois, Garrard Publishing Company, 1966). (The Spache Readability Formula was used with the sample stories presented in this chapter.)

[7]MacGinnis, George H., "The readability graph and informal reading inventories," *The Reading Teacher,* March, 1969.

some cases elementary strategies. Games that involve chasing and fleeing, and putting one small group in competition with another, as well as those involving fundamental skills mentioned above are suggested. The games should be simple enough to learn and they should capitalize upon the imitative and dramatic interests which are typical of this age (this applies to stunt and rhythmic activities as well). Children at the upper elementary level retain an interest in some of the games they played at the primary level, and some of them can be extended and made more difficult to meet the needs and interests of these older children. In addition, games may now be introduced which call for greater bodily control, finer coordination of hands, eyes, and feet, and more strength.

In developing game stories it is also important to strive for maximum activity of all children, avoiding procedures which tend to eliminate players. It may be better to devise some sort of point-scoring system than to eliminate a player from the activity when he is tagged in a running game or hit with the ball in a dodge ball-type game.

In summary, motor-oriented reading content adds variety to the reading program and utilizes still another way of developing concepts in mathematics. As such it could be said that a kind of integration can be realized as far as reading and mathematics are concerned.

## SOME GENERAL GUIDELINES FOR THE USE OF MATERIALS

Whether the materials for mathematics motor activity stories are prepared by the teacher or are professionally prepared, there are certain generalized guidelines which might well be considered for their use. In general, these center around (1) introducing the material, and (2) independent reading and follow-up activities.

### Introducing the Materials

After the prepared stories have been made available for the classroom library, the teacher may introduce several stories by reading them to the children (or tape recording them) and then having the children play the game or demonstrate the stunt or rhythm. Stories developing each type of motor activity should be selected so children will understand how the stories provide details they can use to figure out how to perform a stunt or rhythmic activity or to play a game. Sample stories should also be selected to demonstrate that some stories can be acted out by an individual child and that some require several children to participate in a game or rhythmic activity. This latter aspect of mathematics motor-oriented reading content material utilizes another basic principle of learning, that is, *the child should be given the opportunity to share cooperatively in learning experiences with his classmates* (under the guidance of the teacher). The point that should be emphasized here is that, although learning may be an individual matter, it is likely to take place best in a group. This is to say that children learn individually but that socialization should be retained. Moreover, sharing in group activities is essential in educating for democracy.

After the teacher provides auditory input with a sample story the children can be asked to carry out the activity. As the children carry out the activity the teacher *accepts* their efforts. The teacher may provide guidance *only* to the extent it is necessary to help the children identify problems and provide opportunities for them to exercise judgment in solving them and obtaining the goal, that of playing the game or performing the stunt or rhythm, and internalizing the mathematical concept inherent in the story. Part of the story might be reread by the teacher if the children have difficulty in understanding how to carry out the activity. The children might be encouraged to discuss ways they could help themselves remember the details of the story.

## Independent Reading and Follow-Up Activities

After such an introduction to the stories some children can be encouraged to read them "on their own." The teacher and children might plan several procedures for using the stories. Such activities might include the following:

1. A group of children may select and read a story for a physical

education activity.

2. Individual children may select stories involving individual stunts for a physical education period.
3. After reading one of the stories an individual child may elect to act out his favorite stunt before a group of children. (The children might be asked to guess who or what the story describes as well as the number experiences that are inherent in the story.)
4. After reading one of the stories an individual child might get several other children to read the story and participate in playing the game.
5. Children might use a buddy system for reading and acting out stories.

It is likely that in most instances the teacher will need to provide the guidance to help children with an understanding of the mathematical concept inherent in a story. However, in many cases children will be alert to this themselves.

Chapter 6

# LEARNING ABOUT NUMBER AND NUMERATION SYSTEMS THROUGH MOTOR ACTIVITY

OPPORTUNITIES for counting and using numbers abound in many children's motor activity experiences. For example, in tag games the children who are caught can be counted and that number compared to the number of children not caught. In games requiring scoring there are opportunities for counting and recording numbers. In fact, it is difficult to identify any kind of active game situation which does not include the use of numbers.

In this chapter and those which follow, many activities are described. For each activity the mathematical concepts involved are noted so that teachers can more easily locate instructional activities for a given area of content. Furthermore, instructions for making the best use of each activity are included. Some of the activities are especially useful for introducing a mathematical concept. These involve the learner actively and incorporate a dramatization of the concept physically. Others reinforce concepts and skills previously taught. These are activities which provide needed practice in interesting and personally involving situations. Of course, teachers will want to adapt many of the activities so they can be used to develop mathematical concepts and skills other than those cited in the descriptions.

The activities described in this chapter involve basic number and numeration concepts. In Chapter 7, activities are described which involve the operations of arithmetic: addition, subtraction, multiplication, and division. Finally, in Chapter 8 are activities which involve other areas of mathematics, including geometry and measurement.

Teachers will find Appendices A and B, the *List of Activities by Title* and the *List of Activities by Concept,* especially helpful when planning instruction. In the *List of Activities by Concept,* activities are listed for specific areas which are likely to be the focus of instruction.

* * * * *

*Concepts*: Rote counting, forward and backward
Ordinal number ideas
Numeration
*Activity*: Pass Ball Relay

Children divide into teams. The team members line up one behind the other, close enough so they can easily pass a ball overhead to the next child in line. On signal, a ball is passed over each child's head to the end of the line. As children pass the ball overhead, each child calls out the number of his or her position on the team (one, two, three, etc.) until the ball reaches the end of the line. When the last child on the team receives the ball, he calls his number and then passes the ball forward again. The next to the last child calls out his number, and the ball continues to be passed forward to the front of the team. The activity can be varied by passing the ball in different ways, e.g. under the legs, or alternating over and under. The winner is the first team to pass the ball forward and back with correct counting both forwards and backwards.

*Instructional Use*: Children gain skill in rote counting while engaged in this activity. Children *do* need to know the sequence of number names if they are going to be able to use that sequence for rational counting and in their study of arithmetic. The teacher may start the counting at any number, depending on the skills of the group. Teams of ten or fifteen children apply numeration concepts when the counting starts with numbers like 195 or 995.

* * * * *

*Concept*: Rote counting 1-10
*Activity*: Catch A Bird Alive*

Children line up along a start line, then recite the following

*Adapted from George E. Hollister and Agnes G. Gunderson, *Teaching Arithmetic in the Primary Grades*. (Boston, D. C. Heath and Co., 1964), p. 52.

verse, an old favorite:

1, 2, 3, 4, 5
I caught a bird alive;
6, 7, 8, 9, 10,
I let him go again.

As the children count 1-5 they run forward five steps, then as they say the next phrase they pretend to catch a bird in their hands. They run back to their original places counting 6-10, and as they say the last phrase they pretend to let the bird go.

*Instructional Use*: Children gain skill in reciting the number names 1-10 in order, an essential skill in a child's early experiences with arithmetic.

* * * * *

*Concepts*: Rote counting
Counting by tens
Cardinal and ordinal number ideas
*Activity*: Bouncing Relay

The children are divided into several teams. Members of each team stand side by side. The first child bounces a playground ball ten times consecutively, calling out the number of each bounce. When he has counted to ten, he passes the ball to the second child. The second child then bounces the ball ten times, calling out the number of each bounce, but on the tenth bounce he calls out "twenty," for that is the total number of bounces for the team. All team members follow the same pattern (1, 2, 3, . . . 9, 30) until each person has bounced the ball and added ten to reach the correct total. For example, if there are eight children on a team, the last child should end his count with "eighty." At any time a child misses before completing ten bounces, he retrieves the ball and continues his counting. The first team reaching the correct total wins.

*Instructional Use*: Children gain needed skill in counting by both ones and tens. Further, counting with each bounce helps to develop cardinal number concepts, for as a child says "five" he is completing five bounces, and as he calls out a multiple of ten he is stating the total number of bounces for the team.

* * * * *

*Concepts*: Rational counting (0-9)
Numerals (0-9)
*Activity*: Hot Spot

Pieces of paper (cardboard, asphalt tile) with numerals from zero through nine are placed in various spots around the floor or play area; there should be several pieces of paper for each numeral. The teacher has a number of large posters with collections of objects pictured on them, including posters with no objects pictured. (The overhead projector may be used to present the different quantities of objects.) A poster is shown to the class, perhaps with the question "How many?" The children must identify the number of objects on the poster and then run to that numeral on the floor. Any child who is left without a spot gets a point against him. Any child who has less than five points at the end of the period is considered a winner.

*Instructional Use*: Children gain skill in counting a set of objects and identifying the digit that shows how many are in a set. After the game the posters can be put on display around the room with the correct numeral by each one.

* * * * *

*Concepts*: Rational counting (1-6)
Numerals (1-6)
*Activity*: Watch the Numerals

On large sheets of paper write the numerals 1, 2, 3, 4, 5, and 6, one numeral per sheet. Be sure to make the numerals large enough that children can read them while marching around the room. Children start the activity by marching around the room to music in single file. Then, the teacher holds up one of the numerals. If the numeral is two, each child tries to find a partner, and they continue marching in pairs. If the numeral is three, the children march in three's. If a child is not able to become part of a grouping, he goes to the sideline until the next numeral is presented, then he rejoins the scramble to get into a correct grouping.

*Instructional Use*: Children gain skill in recognizing the numerals 1-6 and in forming groups for the number indicated. Be alert for the child who is challenged by having to include himself within the group he is forming.

* * * * *

*Concepts*: Rational counting
Greater than, less than
*Activity*: Bee Sting

Three children are bees; they are in their hives marked in chalk on the play area. The rest of the children are in the center of the designated play area. The bees run out and try to catch (sting) the children. When a child is caught, he must go with the bee to the bee's hive. When all the children are caught, each bee counts those in his hive. Different children should have a chance to be bees.

*Instructional Use*: Children are provided another opportunity to practice rational counting. The relations greater than and less than are applied as the numbers of children in the hives are compared.

* * * * *

*Concept*: Rational counting
*Activity*: Chain Tag

A child is chosen as leader, then the leader chooses another child to assist him. The two join hands, then chase the other children attempting to tag one. When a child is tagged the chain grows, for he takes his place between the other two. The leader and his assistant, the first two children, remain at the ends of the chain throughout the game and are the only ones who are able to tag. Whenever the chain encircles a child, he may not go under the hands or break through the line. If the chain breaks, it must be reunited before tagging begins again. Every time another child is added to the chain, the leader counts out loud to determine how many children are in the chain. When the chain consists of five or ten children, the game ends. A new leader is then chosen and the game continues in like manner.

*Instructional Use*: Children gain skill in rational counting as they determine the number of children in the chain. The sequence of number names becomes more automatic and the cardinal use of numbers is reinforced.

* * * * *

*Concept*: Rational counting
*Activity*: Round Up[1]

All of the children except ten take places in a scattered formation within the play area. The ten children join hands in a line, for they are the Round-Up Crew. The other children are the Steers. On a signal the Round-Up Crew, with hands joined, chases the Steers and attempts to encircle one or more of them. To capture a Steer, the two end children of the Round-Up Crew join hands. As Steers are captured, the children count the number aloud. It may be necessary for them to stop and count each time another child is caught. When ten children have been captured, they become the Round-Up Crew, and the game continues.

*Instructional Use*: In this game children match number names in sequence to definite objects, in this case, children. Practice in rational counting is provided as children check the number of Steers caught.

* * * * *

*Concept*: Rational counting
*Activity*: Fish Net

After the class is divided into two teams, each team chooses a captain. One team is the Net; the other is the Fish. As the game starts, the teams stand behind two goal lines at opposite ends of the playground, facing each other. When the teacher gives a signal, both teams run toward the center. The Net holds hands and tries to catch as many Fish as possible by encircling them. The Fish try to get out of the opening before the Net closes. They are not permitted to go through the Net by going under the arms of the children, but if the Net breaks because children let go of each other's hands, then the Fish can go through that opening until hands are joined again. The Fish are safe if they get to the opposite goal line without being caught in the Net. When the Net has completed its circle, the captain counts the number of Fish inside. The next time, the teams change places. The team with the

---

[1]Humphrey, James H., *Child Learning Through Elementary School Physical Education.* (Dubuque, Iowa, Wm. C. Brown Co., 1966), p. 194.

largest catch wins the game.

If it is feasible to present a problem situation requiring addition, have the captains record the number of their catches and continue the game until both teams have had a chance to be the Net three times. Then the team with the highest score wins.

*Instructional Use*: Children gain skill in counting to find how many are in a set and, when feasible, recording numerals and adding. Rational counting, i.e. enumeration, rather than rote counting is involved because children find the number name for the numerousness of the entire collection.

* * * * *

*Concepts*: Rational counting
Addition of whole numbers
*Activity*: Card Toss

Several teams are formed from the class, and a set of ten cards is made for each team. Each team member tosses the ten cards singly into a wastebasket, a wide-mouth can, or a propped-up hat about ten feet away. Each card successfully tossed scores one point for the team. After a child has tossed all ten cards, he counts the number of cards within the container and reports the score to his team's scorekeeper. Younger children may need to be helped by the teacher in order to add the total score; however, children will eventually be able to add their own team's total score.

*Instructional Use*: Children gain skill in counting and in simple addition within a highly motivated situation. Encourage children to check their own counting as well as that of other members of their team in order to give maximal opportunity for all to develop skill in rational counting.

* * * * *

*Concepts*: Rational counting
Reading numerals (0-9)
Zero as the number of the empty set
*Activity*: Ball Bounce

Prepare ten large sheets of paper by numbering them from zero through nine. Arrange them in mixed order on the floor about fifteen inches apart, and tape them securely to the floor. Have the children sit in a circle and let them take turns bouncing a ball on each sheet of paper the number of times indicated by the numeral on the paper. When a child does the complete set correctly he scores a point. Other children can count as the child bounces the ball to see if the ball is bounced the correct number of times.

*Instructional Use*: Children gain skill in rational counting in a situation which requires matching number names with nonconcrete sets (sets of bounces) — a much-needed variation, as most instruction involves concrete sets. Children also gain practice reading one-digit numerals from a variety of positions. By including a sheet of paper with the numeral zero, children reinforce the idea that zero is the name for the number property of empty sets.

* * * * *

*Concepts*: Rational counting
*Activity*: Ball Pass[2]

The children are divided into two teams, and both teams form one single circle. If the group is large, the teacher may have two circles with two teams in each circle. The teacher gives directions for a ball to be passed or tossed from one child to another. The teacher may call, "Pass the ball to the right," "Toss the ball to the left over two children," "Toss the ball to the right over four children," varying the calls by the numbers and directions given. The game may be made more complex by using more than one ball of different sizes and weights. If a child drops a ball, a point is scored against his team. The team with the lowest score wins.

*Instructional Use*: Children gain skill in rational counting. They also practice identifying right and left.

* * * * *

*Concepts*: Rational counting
Greater than, less than
*Activity*: Mrs. Brown's Mouse Trap[3]

---

[2]Humphrey, James H., *Child Learning Through Elementary School Physical Education*. (Dubuque, Iowa, Wm. C. Brown Co., 1966), p. 129.
[3]Humphrey, James H., Teaching Children Mathematics Through Games, Rhythms, and Stunts. LP No. 5000, Deal, N. J., Kimbo Records, 1968.

Some of the children stand in a circle, facing in. The children hold hands up high to represent a mouse trap. The other children are Mice. On a signal the Mice go in and out of the circle. One child is Mrs. Brown. When she calls out "Snap!" the children in the circle drop hands (still joined). As the mouse trap closes, some of the Mice are caught in the circle (the trap). The teacher helps the children count how many Mice were caught and how many were not caught. The children should exchange places for the next game.

*Instructional Use*: Children practice matching number names one-to-one with objects as they count to find how many children were caught and how many were not caught. Not only can the number of children in each group be determined, but the groups can be compared. Do they have the same number? Which is greater? Which is less?

* * * * *

*Concepts*: Rational counting (1-9)
Identifying numerals and number words for a given set (0-9)
Reading numerals and number words (0-9)
Zero as the number of the empty set

*Activity*: Show-Me-Relay

Teams of three children each participate in this relay. The teams line up behind a start line, and a chair is placed at a distance for each team. The first team member has a bag of objects, e.g. beans, plastic discs, small cubes; the second has ten numeral cards, one each for the digits 0 through 9; and the third child has number word cards, one each for the numbers zero through nine.

The leader asks the children to show one of the numbers zero through nine. If he calls out "Show me five," the first child races to the chair, places five objects on the chair, and returns to touch the hand of the second child. The second child places the numeral card for five on the chair, and the third child places the number word card for five on the chair. A team is finished when the third child returns to the start line. Whenever the leader calls "Show me zero," the first child touches the chair but, of course, does not leave any objects on it. The first team to finish and have the correct materials on the chair gets one point. A team which finishes first but does not have the correct materials on the chair looses a point. The first team to get five points wins the game. Roles of the three team members should be exchanged from time to time.

*Instructional Use*: Children gain skill in matching sets with the numerals and number words which show how many are in a set. If children are permitted to assist the leader in verifying that the winning team has placed the correct material on their chair, practice is also provided in rational counting and oral reading of both numerals and number words for zero through nine.

* * * * *

*Concepts*: Rational counting
Pairs of addends for ten
Missing addends
Column addition

*Activity*: Count the Pins

Plastic bowling pins or milk cartons are set up at one end of the playing area. These pins are arranged with four on the back row, three on the next row, two on the next, and one on the front row. When set up, they form a triangle. Each child has two turns to roll a plastic bowling ball. He stands behind a line to roll. The total number of pins knocked down in these two rolls is his score. Each child may have several turns in a game, and his total score is determined.

*Instructional Use*: Children gain skill in rational counting as they count the number of pins knocked down. For variety, have the children count the number of pins which remain standing and apply their knowledge of addend pairs for ten in determining how many are knocked down. For example, if four remain standing, say "There are ten pins altogether. If one part of ten is four, what is the other part? Yes, four and *six* is ten." If possible, children should be encouraged to add their scores to obtain their total game score, thereby gaining experience with column addition.

* * * * *

*Concepts*: Rational counting
Reading numerals
*Activity*: Count and Go

The children line up along the long side of a rectangular hard surface court. There are parallel lines drawn in chalk on the court. These are unequal distances apart, and parallel to the long side of the court. The teacher stands across from the children with numerals cards. As the teacher holds up any card at random (numerals on the cards are from one to the number of chalk lines which are drawn), the children must count the lines as they run, skip, etc. toward the teacher. When the children have progressed as many lines as indicated by the numeral card, they stop and stand still. The child who reaches the far side first is the winner. To keep the group small, the girls might take a turn and the boys watch to see if they count correctly. Then they can change places with the boys for the next game.

*Instructional Use*: Reading numerals and rational counting is reinforced as children move forward the varying number of lines indicated on the card. The activity can be varied by relating the direction of a move to the operations of addition and subtraction, as is often done with a number line. Directions, such as *plus two* and *minus three*, could be presented, and the children would then proceed to carry out these directions in terms of moving forward or backward.

* * * * *

*Concepts*: Greater than, less than
Ordinal number ideas
Even and odd numbers
*Activity*: Number man

One child, the Number Man, faces the class which is standing on a line at the end of the play area. Each child in the line is given a number by counting off. The Number Man calls out, "All numbers greater than ____." The children who have numbers greater than the one called must try to get to the other side of the play area without being tagged by the Number Man. The Number Man may also call out, "All numbers less than ____," "All even numbers," or "All odd numbers." Anyone who is tagged must help the

Number Man tag the runners. Any child who runs out of turn is considered tagged.

*Instructional Use*: Children gain skill with number sequence while identifying numbers which are greater or less than a given number. When children are lined up in sequence, it can be observed that the sets of odd and even numbers involve every other whole number.

* * * * *

*Concepts*: Ordinal number ideas
Multiples of three
*Activity*: Leader Ball

Two teams stand in circle formation. On a given signal the leader of each team passes a ball to the player on his right, who passes it to the next player, and so on until it reaches the leader. The leader calls, "first round" immediately and continues to pass the ball for the "second round" and "third round." At the end of the third round, the leader raises the ball to signify that his team has finished. A point is scored for the team finishing first.

*Instructional Use*: The time interval between "first round," "second round," and "third round" requires that children keep in mind which number comes next. A variation requires that after three rounds of play the ball is raised and a point is awarded, as above; however, the numbering of the rounds continues successively. During the second period of play, the leaders count "fourth round, fifth round, sixth round." Attention is drawn to numbers which are multiples of three.

* * * * *

*Concept*: Ordinal number ideas
*Activity*: Number Race

The class is divided into three teams of ten each. In each team, members are assigned numbers from one to ten. They line up behind a starting line in correct numerical order. When the leader gives a signal, the teams race to the finish line and sit down one behind the other in correct number order and as teams. The first team finished scores a point. Number assignment should be changed frequently.

*Instructional Use*: Children are helped to note which numbers come before and which come after their own number assignments. Be sure to change number assignments several times in order to give the children greater facility with the ordinal use of numbers.

* * * * *

*Concept*: Ordinal number ideas
*Activity*: Fetch and Carry

Two lines are drawn about 25 feet apart, one a starting line and the other the goal line. The class is divided into teams of six. If the group is large, several teams will be playing. Each member of the team is assigned a position: first, second, through sixth. It may be helpful for some children if the teacher has each child, as he stands in line, call out the number of his position. The object of the game is for each team to get all its members from the starting line to the goal line. The teams line up at the starting line, facing the goal line. On a signal the first child on each team calls out "first," takes the hand of the second child, and runs with him to the goal line. The first child remains there. The second child runs back to the team, calls out his number, "second," and takes the hand of the third child on the team. They run to the goal line. Now, the second child remains there while the third child returns to get the fourth team member. This procedure continues until one team wins by getting all of its members across the goal line and in correct order first. (It may be wise for the teacher to walk through the relay procedures with the children so they know what to do. After this "practice game," play can begin.)

*Instructional Use*: Children gain increased experience with numbers used in order, including the special number names which can be used in such contexts.

* * * * *

*Concepts*: Ordinal number ideas
Multiples
*Activity*: Call and Catch[4]

---

[4]Adapted from Humphrey, James H., "Teaching Children Mathematics Through Games, Rhythms, and Stunts." LP No. 5000, Deal, N. J., Kimbo Records, 1968.

Children stand in a circle, and each child is assigned a different number. The teacher throws the ball into the air and calls out a number by saying, "Just before six" or "Next after five." For example, if the teacher calls "Next after five," the child assigned number six tries to catch the ball after it bounces. A variation involves assigning multiples of a number to the children. If children are assigned multiples of five and the teacher calls "The multiple of five just before 30," the child assigned 25 tries to catch the ball after it bounces.

*Instructional Use*: Children gain skill in using numbers in sequence, and the variation provides practice recalling multiples of a specified number. The teacher can provide for individual differences of children through appropriate pacing. For example, for the slower child the teacher can call the number and then momentarily hold the ball before throwing it in the air.

* * * * *

*Concepts*: Successive doubling
Grouping pattern for Hindu-Arabic numeration (base two)
Powers of two

*Activity*: Muffin Man

Children stand in a circle. While all the children sing the question of the song "Muffin Man," two children stand still in the center of the circle where they place their hands on their hips and face each other. After the question is sung, the two children in the center clasp hands and skip around the inside of the circle as everyone sings the answer: "two of us know the Muffin Man . . ." and so on. When they finish the answer, the two children stand in front of two new children. Everyone sings the question again, then the four children in the center clasp hands and skip around the inside of the circle as all the children sing the appropriate answer: "Four of us know the Muffin Man . . ." and so on. This procedure is repeated for eight, sixteen, and so on, depending on the size of the group.

Question: Oh, do you know the Muffin Man,
the Muffin Man, the Muffin Man?
Oh, do you know the Muffin Man
who lives in Drury Lane?

Answer: Two of us know the Muffin Man,
the Muffin Man, the Muffin Man.
Two of us know the Muffin Man
who lives in Drury Lane.

*Instructional Use*: Children gain experience with successive doubling through this rhythmic activity. The grouping pattern associated with base two numeration is also illustrated. Number sentences showing the doubling which occurs in the game might be written on the chalkboard, both addition and multiplication number sentences. Exponential notation could even be introduced.

| | | |
|---|---|---|
| $1 + 1 = 2$ | $2 \times 1 = 2$ | $2 = 2^1$ |
| $2 + 2 = 4$ | $2 \times 2 = 4$ | $2 \times 2 = 2^2$ |
| $4 + 4 = 8$ | $2 \times 4 = 8$ | $2 \times 2 \times 2 = 2^3$ |
| $8 + 8 = 16$ | $2 \times 8 = 16$ | $2 \times 2 \times 2 \times 2 = 2^4$ |

* * * * *

*Concept*: Counting by tens
*Activity*: Red Light

Two parallel lines are marked off about thirty feet apart in a play area. One child is *It*. The child who is *It* stands on one line, the goal line. The remaining children stand along the other, the starting line. *It* turns his back to the children and counts loudly, "10, 20, 30, . . . 100, Red Light!" Children advance toward the goal line as he counts, but they must stop as he calls "Red Light." As *It* calls "Red Light" he turns, and if he sees anyone moving, he sends the child back to the starting line. The object of the game is to see which child can reach the goal line first.

*Instructional Use*: Practice in counting by tens is provided for all children participating. They have to anticipate the position of one hundred in the sequence in order not to be caught off guard.

* * * * *

*Concepts*: Roman numerals
Ordinal number ideas
Multiples
*Activity*: Roman Numeral Relay

The children are divided into teams. Cards are made for each team with Roman numerals from I to X (later to XX). The cards are mixed up and placed in a box a specified distance from the starting line for each team. The first child of each team runs to the team box and looks for the Roman numeral "I" and returns to his team where he places the card along the starting line. The second child then runs to the box to find the Roman numeral "II," runs back and places the numeral beside the "I" along the starting line. The third child then runs to the box. This continues until the entire set of numeral cards has been carried to the starting line. If the teams are small, children may run two or more times before all of the cards have been moved. The first team to have all of the numeral cards along the starting line in correct sequence wins. To vary the game with older children, use a set of numeral cards for multiples of X, C, or M.

*Instructional Use*: This activity provides reinforcement of the ability to recognize Roman numerals. While playing the game, children also practice using numbers in sequence. The members of each team should be reordered between games.

* * * * *

*Concepts*: Roman numerals
Cardinal number ideas
*Activity*: Roman Numeral Bounce

The game starts with everyone standing along a line some distance away from and parallel to a chalkboard. The children count one through ten (later eleven through twenty) until everyone has a number. The teacher then bounces out a number. The children who have that number race to the chalkboard to write the Roman numeral for the number of times the ball was bounced. The first one to write the numeral correctly gets to choose the next number and bounce the ball.

*Instructional Use*: This activity is designed to reinforce previous instruction in Roman numeration. As children play the game, they count as the ball is bounced and record a numeral as the name for the number of bounces altogether.

* * * * *

*Concepts*: Number Line
Identifying numerals
*Activity*: Number Line Relay

The class is divided into teams of ten. A set of numeral cards from one to ten is provided for each team. Also, a line to be used as a number line is drawn on the chalkboard in front of each team. The line should be about ten inches above the chalk tray and have dots clearly marked at equidistant points. One of the points should be labeled with a zero under it, with at least ten unlabeled points to the right. On a signal the first child on each team selects from the team's cards the numeral for the number the teacher calls out. When the child has found the card, he runs up to the team's number line and places the numeral card under the point for the number named. When the card is in place, he returns to his team and the child behind him selects the numeral for either the number which would follow the first number called *or* the number which would come before the number called. Each child on the team proceeds in this manner until all the numeral cards are correctly placed on the number line. The first team finished wins.

*Instructional Use*: Experience with a number line is provided in an activity requiring active involvement on the part of every child. Before children can be expected to enjoy success in this activity, they should be able to count to ten and identify each numeral when given the oral number name. Many variations are possible: sets of numerals cards for multiples of ten (10, 20, . . ., 100), multiples of one hundred, or multiples of 5; a set of cards for integers, including a few negative numbers (the zero should be positioned nearer the center); or possibly a set of fractions.

* * * * *

*Concepts*: Numeration
Reading numerals greater than 100
*Activity*: Postman Game

The class is divided into two teams. Members of one team are the postmen and are given envelopes, each having a house number written on it. Members of the other team represent houses and hold numeral cards in their hands. Each postman runs to the various houses trying to find the correct address so he can deliver his envelope. Also, each postman must be able to read the numeral aloud. When all the letters have been delivered to the houses, the teams exchange places. The team that delivers the mail in the shortest period of time wins.

*Instructional Use*: This game is a reinforcement activity which helps children read numerals above one hundred quickly. The game can be varied by using any numerals that the teacher wishes to have the children practice reading.

* * * * *

*Concept*: Multidigit numerals for whole numbers
*Activity*: Find Your Place[5]

Two or more ten-member teams are formed. Ten cards with the digits 0 through 9 are given to each team, one card to each child. The teams are in back of the start line as the teacher calls out a multidigit number name such as "583." On each team, children with the digit cards needed to show that number run to the show line. There they arrange themselves facing the other children and place the numeral cards in front of them so that the specified multidigit number name is clearly readable. The team which shows the numeral first gets a point. The first team to get ten points wins the game.

*Instructional Use*: Children gain skill in translating an oral number name into a numeral, and develop increased awareness of place values (i.e. the powers of ten assigned to each position in whole number numerals). Introduce numerals with zeros gradually, for some of these will be more challenging. For advanced groups, vary the game by asking for the "shortest name" for 200 + 160 + 14. Be careful not to call for a number using any given digit more than once.

* * * * *

*Concept*: Numeration
*Activity*: Place Value Relay

---

[5]Adapted from Pearson, James R., A favorable learning environment, *The Slow Learner in Mathematics,* Thirty-fifth yearbook of the National Council of Teachers of Mathematics, Washington, 1972, p. 115.

The class is divided into two teams. Toothpicks, popsicle sticks, or other appropriate items are available as ones, bundles of ten, and bundles of ten tens (or bundles of a hundred). Or, base ten blocks can be used if they are available in ample supply. The teacher names a number, such as 72. Each team must figure how many bundles of ten and how many ones are needed to show a collection for the number named. The first child runs from the starting line and picks up one bundle of ten (or one "long" if base ten blocks are being used) and brings it back to the team. The second child runs and gets a bundle of ten, and this procedure is followed until the seventh child gets the last bundle of ten needed. The eighth and ninth child then run and each gets one in order to show 72. The first team with the correct display for the number named gets one point. It is important for children to understand that they pick up only one item at a time, regardless of size. For the next number named by the teacher, the child who would be next in line starts the running to get the item decided upon by the team.

*Instructional Use*: Children learn to evaluate a given numeral in terms of the place value (ten and one) and the total value (seventy and two) associated with each digit. They have the opportunity to use numeration concepts in an interesting and highly motivating activity. Teachers may find it helpful to display the numeral as it is called so children can associate each digit with its position in the numeral.

* * * * *

*Concept*: Meaning of fractions (part of a unit region)
*Activity*: Flannelboard Fractions

The children are divided into teams. Flannelboards are set up, one for each team. On display on each flannelboard is a unit region (probably a square region, a circular region, or a rectangular region). About fifteen feet away each team stands in a row and is given a kit of fractional parts prepared for use with a flannelboard. On a given signal the teacher calls out a fraction. The first child for each team selects the correct fractional part(s) from the team kit, runs, places the fractional part(s) on top of the unit region on the flannelboard, and returns to his team. The first child to do this scores a point for his team. The teacher then calls out another fraction, and the second child for each team proceeds in the same manner. The team with the highest score wins. For a variation the teacher can hold up a card with a fraction for the children to show.

*Instructional Use*: Children are enabled to better understand a fraction as part of a unit region as they compare fractional parts and make decisions in the context of a game activity. They also learn by observing the attempts of other children. The teacher can help children find the fraction called for by using the fraction kit and covering a unit region with fractional parts. In every instance, fractional parts should be placed on top of a unit region.

* * * * *

*Concepts*: Meaning of fractions (part of a unit region)
Equivalent fractions
*Activity*: Fraction Target Relay

A diagram is drawn (a circle, square, or rectangle) on the play area, and then marked in sections: one half, one quarter, and two eighths. A throwing line is drawn ten or fifteen feet from the diagram, and teams are formed. Each child on a team gets to throw beanbags until three of them land on the diagram. In scoring, each fractional part is worth a different score. The one-half section is worth four points, the one-fourth section is worth two points, and the one-eighth section is one point. The score-keeper for each team totals the individual scores. The team with the lowest score wins.

*Instructional Use*: When introducing this game, call attention to the fact that the different fractions have varying point values, that one-half is equal to four-eighths so they make four points when the beanbag hits the fractional part one-half, and so on. This game also helps the children to visualize the relationship of the fractions in terms of size.

* * * * *

*Concept*: Meaning of fractions (part of a set)
*Activity*: Corner Spry

A circle about ten feet in diameter is marked in the center of a rectangular playing area. Four teams are formed, each with the same number of members, and a captain for each team is chosen. One team is in each corner of the playing area while the four captains take their places inside the circle. Each captain has a ball. A caller is selected, and a scorer may be chosen as well. Each team member is one of several team members in a corner; that is, he is one part. For example, he may be thought of as one-third, one-fourth, or one-eighth of the team members in the corner depending upon how many there are.

The caller calls different fractions in any order, being careful to use as the denominator the number of team members in a corner. When the caller calls a fraction, that part of the team members in the corner must squat. The captain of each team then passes and receives the ball with each member of his team who remains standing. The first team finished with this exchange without dropping the ball scores a point, and another fraction is called.

*Instructional Use*: Children should be helped to note that the number of children who squat (numerator) is named before the number of team members in a corner (denominator) when saying what part of the group is to squat. If possible, relate the numbers to the written symbol during a discussion of the game.

* * * * *

*Concept*: Meaning of fractions (part of a set)
*Activity*: End Ball

A play area twenty-five by fifty feet is marked off. The area is then divided by lines so that four lines (including the end lines) are equidistant from each other, about twelve feet apart. The class is divided into two teams, and each team is again divided into two groups. The groups from each team take places along the four lines so that they alternate with the groups of the other team. The object of the game is for each team to try to throw the ball over the heads of the opposing team. Points are scored when the ball is caught by the opposing team as it is thrown over their heads. The team with the highest score within a given time limit wins.

*Instructional Use*: Children learn to find half of a set as they are helped to find one-half of the class to make up the two teams. As there are two groups within each team, it can be brought out that each group represents one fourth of the total class. It can also be pointed out that to divide the class into fourths, the total number of children in the class could have been divided by four.

* * * * *

*Concept*: Meaning of fractions (part of a set)
*Activity*: Hit Or Miss

Children are divided into teams. Each team is given an eraser which is set on a table, a chalk tray, or the floor. Teams stand in rows a specified distance from the erasers. Each child in turn is given three or four erasers. He tosses the erasers trying to knock down the target eraser set up on the table. After each child finishes his play he calls out his score, expressing it as a fraction. One hit in four tries is one-fourth; three-fourths would represent three hits in four attempts, and so on. The team with the highest score wins.

*Instructional Use*: Children are enabled to better understand fractions as they use them to name the part of their attempts which were successful. This activity can easily be adapted to those situations in which children practice skills such as shooting baskets or any other type of throw for accuracy. A more advanced variation requires that, after every throw, each child states the fraction for the attempts which were successful for the entire team up to that point in time. If each child had four erasers, the first child would express fractions which include halves, thirds, and fourths. The second child's first throw would be followed by a fraction in fifths because it would be the fifth attempt on the part of the team.

* * * * *

*Concept*: Decimals
*Activity*: Roving Decimal Point

Children line up in teams with about five children to a team. Members of a team are side to side, possibly facing a line marked on the playing surface. A team captain stands in front of each team ready to record scores. A ball, representing the decimal point, is given to a child at the end of each team. On a given signal the ball is bounced or passed from child to child up the line and then back again. When the stop signal is given, the decimal point (the ball) stops and is put to the left of the child last holding the ball. Each child then has to tell what place value he represents. He responds as if he were facing a decimal, that is, if the point is to his right his place value is one or greater, and if the point is to his left his place value is less than one. A point is given for each correct answer. If the ball is dropped, it must be taken to the first child and started down the line again.

When everyone has learned the value of each place to the right and to the left of the decimal point, the children might be assigned numbers and, if possible, given numeral cards. The game proceeds as before, except this time the captain is required to read the multidigit numeral he faces in order to score. The captain can join the line, and the last child becomes captain after each scoring until each has a turn at being captain. The team having the highest score wins.

*Instructional Use*: This game facilitates understanding and skill in naming place values less than one and in reading decimals.

Chapter 7

# LEARNING ABOUT THE OPERATIONS OF ARITHMETIC THROUGH MOTOR ACTIVITY

MOTOR activities can provide children with valuable experiences with the operations of arithmetic (addition, subtraction, multiplication, and division). The energetic involvement of children in such activities brings an interest and enthusiasm to the learning of arithmetic that many children need very much.

In this chapter, activities are described which incorporate the operations of arithmetic. Some activities involve the child with the meaning of the operation, and other activities include computing. For example, addition and subtraction are used in many games which require scoring. For a given operation, activities focusing on the meaning of the operation should be incorporated into the instructional program before activities including computation.

In order to make it easier to find instructional activities for a given operation, the mathematical concepts involved in each are listed with the description of the activity. Also, each description is followed by a discussion designed to help teachers make the best possible use of the instructional activity. Teachers will want to be alert to opportunities to further extend the mathematical experiences of children with these activities by keeping records, charts, and graphs, and by developing concepts of average and percentage.

The reader will find some of the activities useful for introducing a concept. They involve the child actively and tend to incorporate a dramatization of the concept physically. Many of the activities provide reinforcement for concepts and skills which have been introduced earlier. Children find such practice interesting, and they become eager participants in such learning experiences.

Teachers should feel free to adapt an activity, making it more appropriate for the developmental level of their students. Often, by merely substituting larger or smaller numbers, an activity can be made useful for a specific group of children.

The *List of Activities by Concept* will greatly assist teachers in finding activities for a specific area of content. It is found in Appendix A of this book.

* * * * *

*Concepts*: Addend pairs for a given sum
*Activity*: Addition Name Hunt

Pieces of paper (cardboard, asphalt tile) with addition combinations are placed in various spots around the floor or play area. Each combination or pair of addends should be expressed in the form a + b, for example, 4 + 3, or 5 + 7. There should be several pieces of paper with different combinations for each of the sums two through twelve (or higher). The teacher has large numeral cards for two through twelve, or for the numbers being presented. The leader says, "Can you find a name for . . . ?" then shows a numeral card to the class. The children run to an addition combination on the floor which is a name for the same number being shown by the leader. For example, if the leader shows "6", children will run to the combinations 2 + 4, 5 + 1, etc. Any child who is left without a correct name for the number gets a point against him. Any child who has less than five points at the end of the period is considered a winner.

*Instructional Use*: Children gain skill in recalling and identifying pairs of addends for a given sum, an extremely important skill for the development of problem-solving ability.

* * * * *

*Concept*: Meaning of addition
Number Sentences
Equals
*Activity*: Lions and Hunters

Two teams are established. One team, the hunters, begins by forming a large circle; the other team, the lions, is within the circle. The hunters use a ball and attempt to hit as many lions as possible within a two-minute period. As the lions are hit, they go to a lion cage that has been marked in chalk on the court. When the teams change places, the second group of lions hit go to a second lion cage marked on the court. A scorer records the number of lions by each cage and completes the appropriate number sentence and labels as illustrated:

| (cage) | | (cage) | | |
|---|---|---|---|---|
| 3 | + | 4 | = | 7 |
| addend | plus | addend | equals | sum |

*Instructional Use*: Involve all of the children in counting the lions in each cage and in counting the number of lions caught altogether. Emphasize the fact that "three plus four" tells how many lions were caught altogether, and "seven" also tells how many lions were caught. They are both names for the same number, and that is what "equals" means; it means "is the same as." Three plus four *is the same as* seven.

* * * * *

*Concepts*: Addend pairs for a given sum
Basic addition facts
*Activity*: Fast Facts

Children are grouped according to a specific number to be thought of as a sum. If the class has been studying pairs of addends for the number eight, children are organized into teams of eight. As the activity begins, children are lined up along a line on the floor or playground, and clustered in their teams. Opposite each team on a parallel line is placed a marker or partition. The leader calls out "Eight equals five plus three," and all teams run to the other line. The children on a team arrange themselves so that five children are on one side of the marker and three are on the other side. When all have agreed that the addition fact is correctly pictured, the leader calls out "Five plus three equals eight," and all the children run back to the starting line and form their teams again. The activity continues as the leader calls out other addition facts for the same sum. If desired, a team can be given a point for being the first to picture the two addends.

*Instructional Use*: Children develop an understanding of the meaning of addition as they associate the partitioning of a set with various pairs of addends for a given sum. As children interpret the orally given number sentences, they become more comfortable with the sum placed before the equal sign. A variation is to present the number sentences visually on large cards or with an overhead projector, possibly just after an attention-getting signal. Be alert for children who may have difficulty including themselves when counting the number in each part of the partitioned group of children.

* * * * *

*Concepts*: Addend pairs for a given sum
Missing addends
Basic subtraction facts
*Activity*: Parts of Seven

Children stand spread out in a circle with the leader in the center. The leader, who has a ball in hand, says "Seven equals four (bounces the ball four times) and . . ." He then tosses the ball to a child who catches the ball and bounces it three times and says "Three" (because $7 = 4 + 3$). If the leader says "Seven equals two . . ." and bounces the ball two times, the child who receives the ball bounces it five times, and so forth. When the activity is varied in order to show two addends for a different sum, the name of the activity should be changed to suggest the sum involved.

*Instructional Use*: Children practice recalling pairs of addends for a given sum as they determine the missing addend. The leader can match the difficulty of specific tasks to the capabilities of the child receiving the ball.

* * * * *

*Concepts*: Basic addition facts
Higher decade addition
*Activity*: Beanbag Toss

A board is made three or four feet square, with a small eight-inch circle cut out in the center of the board. A child standing ten feet from the board tosses five beanbags, one at a time, at the board. If he hits the square board he scores one point. If the beanbag goes through the hole in the center he scores three points. After throwing the five beanbags, a child determines his score for that turn and records it. A cumulative total is also computed. The child with the highest score wins.

*Instructional Use*: This activity enables children to use their addition facts and skill with higher decade addition in an exciting and highly motivating situation. For more advanced children, assign larger numbers for scoring purposes.

* * * * *

*Concept*: Addition
*Activity*: Addition Beanbag Throw

Five large-mouth cans are tied together and are numbered from one to five. The class is divided into teams, and a set of cans is provided for each team. The teams stand in rows behind a line about ten feet from the targets. Each team member throws a beanbag, trying to get it in can number five, as this is worth the most points. Each child has three tries. At the end of his turn, each child adds up his own score, and it is recorded for his team. When all the children have had their turns, they add the scores to find which team has the most points. This team wins. In case of a tie, there is a play-off.

*Instructional Use*: A game of this type stimulates interest in arithmetic reinforcement activities. Play the game with higher numbers as the children progress. Or, concentrate on specific numbers. For example, if the group has been working with the set of addition and subtraction facts with four as an addend, label two or three of the cans with a four.

* * * * *

*Concept*: Addition of multidigit numbers
*Activity*: Bean Bag Addition

The class is divided into teams, and for each team a target is drawn on the playing area. Target areas should be about four or five feet square with six circles (each ten inches in diameter) drawn in the square and marked with different three-digit numerals. A child is chosen to serve as scorekeeper. The first player on each team throws three beanbags at the team's target. The numbers for the circles that are hit are written down and added by the child who threw the beanbags. The scorekeeper compares the sums for the first child from each team, and the team with the largest sum gets a point. The second child for each team then throws three beanbags, and so on. The team with the most points wins.

*Instructional Use*: This game provides practice for the skill of adding multidigit numbers in a highly motivating situation. Encourage team members to verify each computation.

* * * * *

*Concepts*: Addition
Subtraction
*Activity*: Add-A-Number Relay

The class is divided into several teams, and a number is recorded for each team on the chalkboard. (Use low, one-digit numerals at first.) The teacher writes or calls out a number, then the first team member of each team runs to the board and adds this number to his team's number on the board. He returns to his team, and the next team member runs to the board and adds the same number to the new sum. Each child on the team does the same until the first team finished wins. Each team should start with different numbers to prevent copying. Size of numbers used will depend on the developmental level of the children.

*Instructional Use*: This game provides reinforcement of addition facts and addition computation presented visually. Wide variation in the difficulty of the addition situation is possible by varying the numbers used. The same activity can be varied by using subtraction. When subtraction is the operation, be sure that the numbers recorded initially on the chalkboard are large enough to allow successive subtractions as anticipated. The subtraction game is called "Subtract-A-Number Relay."

* * * * *

*Concepts*: Basic addition facts
Basic subtraction facts
*Activity*: Number Catch

Every child is given a number from one to ten. The teacher calls "Two plus two" or "Six plus one" and tosses the ball into the air. Any child whose number happens to be the sum of the numbers called can catch the ball. The other children run away as fast as they can until the child catches the ball and calls "Stop." At that time all the children must stop where they are and remain standing in place. The child with the ball may take three long, running strides in any direction toward the children. He then throws the ball, trying to hit one of the children. If he succeeds, the child who is hit has one point scored against him. The game continues, with the teacher calling out another pair of addends. The children with the lowest number of points are the winners.

*Instructional Use*: This reinforcement activity encourages immediate recall rather than just figuring out the sum again. Assign the numbers nine through eighteen if the children have studied the basic addition facts with larger sums. If the game is altered for the operation of subtraction, the teacher calls "Eight minus three" and "Six minus zero."

* * * * *

*Concepts*: Basic addition facts
Basic subtraction facts
*Activity*: Steal the Bacon (Variation)

The class is divided into equivalent teams, and each member is given a number. The teams line up some twenty or twenty-five feet apart, with an object at the center equidistant from both teams. The children on each team should be mixed up so they are *not* lined up in numerical sequence. The regular game of Steal the Bacon is played; however, the teacher calls out a simple addition or subtraction problem. The child from each team whose assigned number is the missing sum or addend must try to steal the bacon first in order to score a point for his team.

*Instructional Use*: This activity is designed to reinforce immediate recall of the basic facts of arithmetic. It is possible to keep in mind the needs of individual children as numbers are assigned and problems are presented. For example, a child assigned the number six who is having difficulty recalling zero facts might be required to respond to 6 - 0 in order to provide practice in an active and emotionally involving context. For variety, use a visual stimulus part of the time.

* * * * *

*Concepts*: Basic addition facts (sums to 10)
Basic subtraction facts (sums to 10)
*Activity*: Trade Places Tag

There are twenty-one children with ten on each team and a tagger. The two teams line up facing each other about twenty feet apart. Each team counts off so each member is assigned a number. The teacher calls out an addition or subtraction problem whose sum is not greater than ten. Children representing the two addends and the sum try to exchange places with children representing the same number from the other team before one of them is tagged by the tagger. For example, for the addition problem 4 + 3 = ?, the two four's exchange places and the two three's likewise. Also, the two seven's trade places. For the subtraction problem 8 - 5 = ?, the eight's, the fives, and the three's attempt to exchange places. If a child is tagged, his team gets one point. The team with the lowest score within a set time wins.

*Instructional Use*: Reinforcement is provided for recalling missing addends and sums for the easier addition and subtraction facts.

* * * * *

*Concepts*: Basic addition facts
Basic subtraction facts
*Activity*: Number Man (Variation)

Each child is assigned a number and stands behind a line at one end of the play area. One child, the Number Man, calls out addition and subtraction problems such as "Five plus six" and "Twelve minus four." The children who have the number which is the answer (the missing sum or addend) for the problem must try to get past the Number Man to the line on the opposite side without being tagged. If tagged, the child must help the Number Man. The teacher will probably want to reassign numbers frequently and have children change places with the Number Man.

*Instructional Use*: Reinforcement is provided for recalling missing addends and sums.

* * * * *

*Concepts*: One less than
Meaning of subtraction
*Activity*: Ten Little Birds

After the children form a circle, ten children are selected for the birds, and they count off from one to ten. They go into the center of the circle and stand in a line within the circle. When the verses are sung, the child in the center with the number being repeated "flies" back to his original position in the circle of children. This is repeated until all the birds have moved back with other children selected to be birds.

Ten little birdies sitting on a line,
One flew away and then there were nine.

Nine little birdies sitting up straight,
One fell down and then there were eight.

Eight little birdies looking up to heaven,
One went away and then there were seven.

Seven little birdies picking up sticks,
One flew away and then there were six.

Six little birdies sitting on a hive,
One got stung and then there were five.

Five little birdies peeping through a door,
One went in and then there were four.

Four little birdies sitting in a tree,
One fell down and then there were three.

Three little birdies looking straight at you.
One went away and then there were two.

Two little birdies sitting in the sun,
One went home and then there was one.

One little birdie left all alone,
He flew away and then there was none.

*Instructional Use*: This rhythmic activity provides children an opportunity to act out the concept of *one less than*. Stop the song at any point and have the children select the vertical and/or horizontal forms of the subtraction fact that records the action that has just taken place. Also, use the phrase *one less than* in the discusion. For example, after the third stanza a child chooses a card with the subtraction fact 8 - 1 = 7 from a set of three cards. As the teacher points to the numerals she says "Eight minus one equals seven. Seven is one less than eight."

* * * * *

*Concepts*: One less than
Meaning of subtraction
*Activity*: Dodge Ball

The class is divided into two teams. One team forms a circle and the second team stands inside the circle. The regular game of Dodge Ball is played, with players in the circle trying to hit the team in the center with a volleyball by throwing it rapidly. The children in the center, to avoid being hit, may move about, jump, stoop, but they may not go outside the circle. When a child in the center is hit, he must go outside the circle. Each time a child is hit, the team forming a circle calls out the subtraction fact for the action which has just taken place. For example, if there are ten children in the center and one is hit, the children call out "Ten minus one equals nine" before throwing the ball again. The team with the largest number of children remaining in the center at the end of two minutes wins the game.

*Instructional Use*: This activity enables children to decide what subtraction facts describe the physical situation they are experiencing. As the number sentences in this activity are all presented orally, the teacher may want to follow with an activity involving the writing of number sentences both horizontally and vertically.

* * * * *

*Concept*: One less than
*Activity*: Musical Chairs

Any number of children form a circle around a set of chairs. There should be one less chair than there are children. As the music plays, children walk, run, hop, or skip around the chairs until the music stops. At this time all children try to find a seat. The extra child is then *out*. Out removes a chair as he leaves the circle of children. The music starts again, and the class repeats as above. The teacher may choose to take out two or three chairs at a time.

*Instructional Use*: In discussing the game with children, note that there was one less chair than children (or two or three less as the case may be).

* * * * *

*Concepts*: One less than
Meaning of subtraction
*Activity*: Ten Little Chickadees

Groups of ten children form lines, facing forward. The children in each line count off one to ten. As the verses are sung, the child with the number for the stanza sits down.

Ten little chickadees standing in a line,
One flew away, now there're nine.

Nine little chickadees standing very straight,
One flew away, now there're eight.

Eight little chickadees looking up to heaven,
One flew away, now there're seven.

Seven little chickadees build a nest of sticks,
One flew away, now there're six.

Six little chickadees looking very alive,
One flew away, now there're five.

Five little chickadees pecking at my door,
One flew away, now there're four.

Four little chickadees very afraid of me,
One flew away, now there're three.

Three little chickadees didn't know what to do,
One flew away, now there're two.

Two little chickadees look toward the sun,
One flew away, now there's one.

One little chickadee hopping on the ground,

He flew away, now there're none around.

*Instructional Use*: As children participate in this rhythmic activity they dramatize the concept of one less than. After the song the teacher may want to have the children write number sentences showing one less, number sentences with "minus one"; for example, 6-1=5. Have children write the sentences in both horizontal and vertical form.

* * * * *

*Concept*: Meaning of subtraction
*Activity*: Tossing Darts

Two circles are drawn on a chalkboard, and a line is drawn at a distance from the board. (The teacher must adjust the size of the circles and the distance for throwing to accommodate individuals.) Two relay teams are counted off. Running to the line, a child from Team A throws from five to ten safety suction-type darts at the first circle. A child from Team B throws the same number of darts at the second circle. The number of darts within the two circles are then compared, and the difference is the score for the team which had the larger number. The second child from each team proceeds in the same manner. The team with the higher score wins.

*Instructional Use*: Children learn that subtraction can be used for comparing two disjoint sets. The process of matching the members of the two sets one-to-one can be demonstrated by removing one dart in the first circle for each dart from the second circle. The remaining darts in the circle with the most darts show the difference. The teacher or a child serving as a recorder may want to write the subtraction number sentences on the chalkboard. Emphasize that in this situation, it makes sense to call the unknown addend a "difference."

* * * * *

*Concepts*: Missing addends
Addend pairs for a given sum
Basic subtraction facts
Higher decade facts
*Activity*: Who Am I?

At least two children should be assigned to represent each number zero through nine. The children stand in a circle facing in, and an object such as a bowling pin is placed in the center of the circle. When the leader makes statements such as "Subtract me from eleven and you have four," children assigned the number seven race to the center of the circle; the first child to pick up the center object gets a point, and the first child to get five points wins the game. From time to time, numbers should be reassigned.

*Instructional Use*: Children gain skill in recalling basic subtraction facts and finding missing addends. By using larger numbers, higher decade facts can be the focus of the activity.

* * * * *

*Concept*: Multiplication by two
*Activity*: Twice as Many

The children stand on a line near the end of the play area and face the caller, who is standing at the finish line some twenty-five to fifty feet away. The caller gives directions such as "Take two hops. Now take twice as many. Take three small steps. Now take twice as many." Directions are varied in number and type of movement. Each direction is followed by "Now take twice as many." The first child to reach the finish line calls out "Twice as many," and everyone runs back to the starting line. The caller tags all those he can before they reach the starting line. All those tagged help the caller the next time.

*Instructional Use*: Children are able to apply their knowledge of multiplication facts for the factor two in a highly motivating activity. The teacher may want to check each time a new direction is given to be sure the children have multiplied by two accurately and have the correct answer. Those children having difficulty could be helped to act out the multiplication fact called for.

* * * * *

*Concept*: Multiplication by two
*Activity*: Grand March

A leader is chosen, and all the children line up behind him and march around the room to marching or walking music. The following pattern is used in marching:

1. March in a circle single file.
2. Follow the leader up the center in single file.
3. The first child goes off to the center, and the second goes off to the left. (Continue with the children going alternately to the right and to the left.)
4. The two new leaders circle and meet at the back of the marching area and come forward as a couple. (All children pair off in the same manner.)
5. The couples come up the center; the first couple goes off to the right, and the second couple goes off to the left. (Continue with partners going alternately to the right and to the left.)
6. The couples circle and meet at the back and form rows of four each.
7. The rows of four come up the center and go off in alternating directions.
8. The rows of four circle and meet, forming rows of eight, and so on.
9. Continue in this manner until all children come forward in a straight line.

*Instructional Use*: When the formation is explained to the children, they might figure out the largest number of children that can be used from their class, given the number of children in the class. They might also find out how many times they will circle during the activity. After marching, the number of sets of two, sets of four, and so on could be determined, with appropriate multiplication and/or division number sentences recorded on a chalkboard.

* * * * *

*Concept*: Multiples of three
*Activity*: Pick-Up Race

A number of wooden blocks (three or more for each child) are scattered over a large playing area. The children divide into several teams and take their place in rows behind a starting line. At the starting line there are circles drawn, one for each team. On a signal the first child runs into the playing area and picks up one block, returns to the starting line and places a block in his team's circle. He then goes back after a second block and returns it to the circle. He gets the third block and leaves the three blocks in a pile in the circle. Play continues in this manner until each child on the team has collected and piled three blocks in the team's circle. The first team that completes the task wins.

*Instructional Use*: While setting up the game, children help determine the number of blocks that will be needed, first by counting by threes or multiplying by three for each team, and then for all of the children. If enough blocks are available, more than three can be placed in each pile.

* * * * *

*Concepts*: Multiples
Multiplication facts
*Activity*: Back to Back

The children stand back to back with arms interlocked at the elbows. The teacher points to each group and, with the help of the children, counts by twos. If one child is left over, the number one is added and the total number of children is thereby determined. The teacher calls for any size group, and on signal the children let go and regroup themselves in groups of the size called for. If the teacher calls for a group of two, the children must find a new partner. Each time the children are regrouped, they count by twos, threes, or whatever is appropriate, and add the number of children left over. (If the resulting number is not the total number of children present, there has been an error and groups should be counted again.) Whenever the number called for is larger than the group already formed, the teacher may choose to ask how many children are needed for each group to become the size group that has just been called for. Whatever the size group called for, the children must hook up back to back in groups of that number. A time limit may be set. The children who are left over may rejoin the group each time there is a call to regroup.

*Instructional Use*: This game not only provides experience with the multiples of a given factor, but also informally prepares children for uneven division. In fact, they may want to predict the number of children who will be left over before the signal to start regrouping is given. If a chalkboard is handy, the teacher may choose to write number sentences to record each regrouping. For example, if there are 25 children and groups of four have been called for, the record should show that six fours and one is 25, or $(6 \times 4) + 1 = 25$.

* * * * *

*Concepts*: Multiples of whole numbers
Common multiples
*Activity*: Multiple Squat

Children stand in a line or a circle, and each is assigned a number in order, starting with one. The children say their numbers in turn. However, whenever their number is a multiple of three they squat but do not speak. A common variation requires that children also squat for any number for which the numeral has the digit three in it. Multiples of different numbers can be used, of course. A more complex variation involves squatting for a multiple of either of two designated numbers, e.g. three and four.

*Instructional Use*: Children develop skill in determining the set of multiples for a specified whole number. The activity can be adapted for more advanced children by having them squat only for common multiples of two numbers. For the numbers two and three they would squat for six, twelve, etc. Children should be reassigned numbers frequently.

* * * * *

*Concepts*: One-digit factors of a whole number
Division with one digit divisors
One as a factor of every whole number
Two as a factor of even numbers
Rules for divisibility
*Activity*: Factor First

Nine children wear or hold cards with one of the numerals 1 through 9. They stand on the outside of a circle facing in, and an object such as a bowling pin is placed in the center of the circle. When the leader calls out a number (such as 24) all children who have a number which is a factor of the number called race to the center of the circle. The first child to pick up the center object gets a point. If a child whose number is not a factor should pick up the object, he looses two points. Play proceeds until the first child gets a predetermined number of points, possibly five points.

It is important that numbers be reassigned rather frequently and regularly. For example, cards can be passed to the person on the left after every third or fourth play.

*Instructional Use*: Children become proficient in finding a missing one-digit factor, a skill necessary for division computation. Some children may also note for the first time that one is a factor of every whole number, or that two is a factor of every whole number with 0, 2, 4, 6, or 8 in the units place. Rules for divisibility can be applied to larger numbers when children have studied them.

* * * * *

*Concept*: Meaning of division (measurement)
*Activity*: Triplet Tag

The children form groups of three, with hands joined. After the groups are formed, the teacher should write a division statement, pointing out that the number of children in the class is the product, and the size of the groups is the known factor. To find the unknown factor the groups are counted. If one or two children are left over, that number is the remainder and it is also recorded. The groups stand scattered about the play area. One group is *It* and carries a red cloth. The *It* group tries to tag another group of three. Hands must be joined at all times. When a group is tagged, it is given the red cloth, and the game continues.

*Instructional Use*: In this game, children act out the measurement meaning of division. By taking a moment to record the numbers in a division statement, children can relate the situation to the symbols they will use when working with paper and pencil.

* * * * *

*Concepts*: Meaning of division (measurement)
Effect of increasing or decreasing the divisor
*Activity*: Birds Fly South

Play begins with the entire class distributed randomly behind a starting line. The number of children in the class is the dividend (or product). A caller gives the signal to play by calling "Birds fly south in flocks of six" (or the largest divisor that will be used). The class runs to another line that has been designated as "South." At this point the children should be grouped in sixes. After observing the number of flocks (the quotient), the children who remain (that is, those who were not able to be included in one of the flocks) become hawks, who take their places between the two lines. Then with the call "Scatter! The hawks are coming!" the children run back to the other line, with the hawks attempting to tag them. Note is taken of who is tagged. Play continues, with the entire class taking its place behind the starting line. The caller then uses the next lower number for the call. If six was used first, five would be called next. "Birds fly south in flocks of five." This continues until groups of two have been formed and they return to the starting line. Each time the children should observe the number of flocks that are formed.

To score the game, each child begins with a score which is the number called for first. In the case illustrated above, the number would be six. If a child is tagged, his score decreases by one point.

*Instructional Use*: At the end of the game, consider the arithmetic which has been applied. If possible, record division number sentences showing the number of flocks formed when different divisors were used for the same dividend. Help the children form the generalization that, for a given dividend (product), when the divisor (known factor) decreases, the quotient (unknown factor) increases in value. After this pattern is established, the numbers called can be reversed beginning with the smallest divisor and working up to the highest divisor to be used. Here, the converse of the previous statement can be developed.

* * * * *

*Concepts*: Meaning of division (measurement)
Effect of increasing or decreasing the dividend
*Activity*: Birds Fly South (Variation)

Starting with the entire class on the starting line, at the signal "Birds fly south in flocks of six" (or the highest number that is to be used as a divisor), the class runs to the line designated as

"South." At this point they should be grouped by sixes or whatever the divisor. After observing the number of flocks (quotient), those children who remain become hawks, who take their places between the two lines. Then, with the call "Scatter, the hawks are coming!" the children ungroup and run back to the starting line, with the hawks attempting to tag them. All children caught, and the hawks, retire to a hawks' nest. A point is scored by the individuals left when the new dividend is less than the divisor being used.

*Instructional Use*: The change in number of children at the starting line (dividend) each time should be noted, as play continues in like manner. Emphasize the number of flocks or groups resulting when the dividend is a smaller number. If a record of division statements can be kept, it will be even easier for children to generalize that for a given divisor (known factor), if the dividend (product) decreases in value the quotient (unknown factor) will also decrease in value.

* * * * *

*Concept*: Basic facts of arithmetic
*Activity*: Call Ball (Variation)

The children stand in a circle, and the teacher stands in the center of the circle with a ball. The teacher calls out a part of the basic fact (7+8, 14-6, 7x7, and 35 ÷ 5 are examples). The teacher then bounces the ball to a child in the circle, and he must try to catch the ball and give the resulting number before the teacher counts to ten. One point is given for recalling the fact, another for catching the ball. The teacher may elect to have a child be the one in the center calling the problems and bouncing the ball. In such a case, it should be emphasized that all children need to be given a chance to catch the ball and give an answer to a problem.

*Instructional Use*: This game is a little easier for children to play than the regular game of *Call Ball*, where they have to be very quick to remember the answer to a given problem. This activity helps to provide needed reinforcement for quick recall of the basic facts.

* * * * *

*Concept*: Basic facts of arithmetic
*Activity*: Cheese and Mice

A round mousetrap is formed by children standing in a circle. In the center of the mousetrap is placed the cheese (a ball or some other object). If subtraction facts are to be practiced, the children are each assigned a number (zero, one, two, up to and including ten). Several children can be assigned to each number. When the teacher calls a number from zero to ten, all the children with the number for the difference between the number called and ten leave their places in the circle, run around the outside of the circle, and then return to their original places (hole in the trap). They then run into the circle to get the cheese. The first child to get the cheese is the winning mouse. The teacher must check to see that the children with the correct number for the answer to the problem were the mice who ran to get the cheese. Only a child who was the correct number can win.

*Instructional Use*: This exciting activity provides an opportunity for children to practice recalling the basic facts of arithmetic in an active situation where each is quite involved personally. As described, the activity focuses on the basic subtraction facts for ten as sum, but the teacher may assign other numbers and call out numbers for more difficult problems according to the skills of the group. Further, the game can be adapted to addition, multiplication, and division. For addition, assign sums to the children and call out pairs of addends. For multiplication, assign products and call out pairs of factors. For division, assign the numbers zero through ten and state that the known factor is 6 (or 7, etc), then call out products of the known factor. For example, if the known factor is 6, and 42 is called, children assigned the number seven would run for the cheese.

* * * * *

*Concepts*: Basic facts of arithmetic
*Activity*: Arithmetic Relay

Children are divided into several teams of equal number, and the teams stand in rows about twenty feet from a chalkboard. The leader calls out one of the basic facts of arithmetic, but with one number missing. "Six plus eight is what number?", "Twelve divided by what number is two?", and "What number minus four equals five?" are examples which might be called. When a fact is called, the first person runs to the chalkboard and writes down the missing number. One point is given to the team who writes down the correct answer first. The game can be played using running, walking, hopping on the right foot, hopping on the left foot, skipping, or jumping with both feet. The teacher may also find it desirable to have the children write the complete number sentence on the chalkboard rather than just the missing number, and they may be asked to write facts both horizontally and vertically. The team with the highest score wins.

*Instructional Use*: This activity provides reinforcement so that children who are learning the basic facts of arithmetic will be able to recall them readily and use them for problem solving. The teacher can provide for individual differences among children while playing the game by carefully selecting the facts to present, and possibly by arranging the children within each team so that children of comparable ability are competing at any given time.

It is easy to adapt this activity for all four of the basic operations: addition, subtraction, multiplication, and division. As illustrated in the examples above, missing numbers can occur at varied positions within the equations. The activity can also be used for computation with larger numbers if the teacher considers it appropriate.

* * * * *

*Concepts*: Basic facts of arithmetic
Greater than and less than
Factors and multiples
Common factors and greatest common factor
Prime numbers

*Activity*: Catch the Thief

The class is divided into two teams of equal number, and members of each team are assigned a number, starting with one for each team. Teams line up twenty or twenty-five feet apart, and an object is placed at the center so that it is equidistant from the two teams. As children line up it is not necessary that they line up in ordinal sequence. Whenever the leader signals, the appropriate child or children from both sides race to the center and try to pick up the object and take it across their line before being tagged by a member of the other team. When successful, one point is scored for the team. The team with the highest score wins.

The choice of appropriate signals for play will depend upon previous experiences of the class. Suggestions include:

"Less than four"
"The sum of five and six"
"The difference between twelve and seven"
"Multiples of four"
"The product of three and five"
"Factors of 24"
"Common factors of eight and twelve"
"Greater than eight and less than twelve"
"Primes"
"The greatest common factor of 12 and 18"

*Instructional Use*: Children reinforce a number of skills, depending upon the choice of signals for play. A call such as "Multiples of zero" will help children generalize the zero property for multiplication as they observe that no children race forward. Also, attention can be directed to the fact that whenever factors for a given number are called for, children with the number one and the given number always get to run. The idea of the empty set is applied with a call like "Greater than five and less than three." To reinforce correct interpretation of similar symbolic (visual) expressions, an overhead projector can be used for presenting the signal.

* * * * *

*Concept*: Addition of fractions (common denominators)
*Activity*: Hit the Target

The class is divided into teams. Each team has a beanbag and a target area drawn on the play area. The target areas should be about five feet square with five circles (each ten inches in diameter) drawn in the square and marked with different fractions, all with the same denominator. Each team also has a scorekeeper. The first player on each team throws a beanbag at the team's target. The fraction within the circle that is hit is recorded by the scorekeeper. After each child has had a turn, the scorekeepers add the fractions; and the team with the highest score wins.

*Instructional Use*: In this type of activity, children have the opportunity to add fractions with a common denominator in a highly motivating situation. Each child on the teams should be encouraged to check the scorekeeper.

* * * * *

*Concept*: Addition of fractions (unlike denominators)
*Activity*: Hit the Target (variation)

The class is divided into teams, and each team has a target area drawn on the play area. The target areas should be about five feet square with seven circles (each ten inches in diameter) drawn in the square and marked with different fractions, all with different denominators. Each team also has a scorekeeper. The first player on each team throws four beanbags at the team's target. The fractions within the circles that are hit are added by the scorekeepers. The sums for the first child from each team are compared, and the team with the larger fraction gets a point. The second child for each team then throws four beanbags, and so on. The team with the most points wins.

*Instructional Use*: Children practice adding fractions with unlike denominators in a highly motivating situation. Each child should be encouraged to work with the scorekeeper as his fractions are being added.

* * * * *

*Concept*: Subtraction of fractions (common denominators)
*Activity*: Train Dodge

The class forms a circle with four children in the center. Children in the center hold each other about the waist, thus making a train. The object of the game is to tag the Caboose, or last child. One or more dodge balls may be used to try to hit the Caboose. As the Caboose is hit, he can wait outside the circle. The game continues until all four children of the train have been hit. At the outset the teacher establishes the concept that the four children within the circle make up the *whole* train, that each child is one part, one-fourth of the whole train. As each child is hit and moves to the outside of the circle, a brief pause is taken to ascertain that one-fourth of the train is now uncoupled and waiting outside and three-fourths of the cars remain coupled together and in the game. When the next child is hit, two-fourths are in and two-fourths are out. Finally, the whole train is assembled outside the ring and is now in the roundhouse. A new train is then formed with another group of four children.

*Instructional Use*: This active game helps children understand subtraction of fractions with common denominators. Midway through the game, the teacher might help the children note that two-fourths are one-half the train and that an equal number of cars are inside and outside the circle. For a variation, use a train of six children and the fractions associated with it.

* * * * *

*Concept*: Multiplication of fractions
*Activity*: Fraction Race

The class is divided into a number of teams. Members of each team take a sitting position, one behind the other a specified distance from a goal line. Starting with the first child on each team, each child is assigned a number (one, two, etc.). The teacher calls out a multiplication combination including a fraction and a whole number. The children whose number is the product stand, run to the goal line, and return to their original sitting position. If the teacher calls out "one-fourth times eight," all the number two's would run. Similarly, if the teacher calls "two-thirds times twelve," all the number eight's would run. The first child back scores five points for his team, the second child back scores three points, and the last child scores one point.

*Instructional Use*: This active game helps children to determine the product of a fraction and a whole number quickly and accurately. Adjust the difficulty of the examples to the experience of the children.

* * * * *

*Concept*: Addition of decimals
*Acitivty*: Dash Relay

A 25-yard course is marked off on the playground. The children are grouped into several teams. At a signal to "Go!" the first child on each team runs from the starting line to the finish line. A timer for each team records the running time with a stop watch. Children place their running time on a team chart, recording it in decimal form. For example, 5.8 seconds might be recorded. When all children have run, the total times for each team are added. The team with the lowest running time wins. The course may be run as often as possible over several weeks as teams try to improve their time.

*Instructional Use*: In this highly motivating situation, children have an opportunity to practice adding decimals. Improvement can be determined by subtracting decimals. Comparisons with other teams and averages may also be investigated.

Chapter 8

# LEARNING ABOUT OTHER AREAS OF MATHEMATICS THROUGH MOTOR ACTIVITY

THUS far, number, numeration, and the operations of arithmetic have been the only mathematical concepts and skills this book has related to learning through motor — or game — activities. Children can, in fact, come to understand a variety of other concepts through this same style of active participation.

Included in this chapter are activities using concepts from geometries of one, two, and three dimensions. Several of these activities will facilitate instruction in measurement, and there are even activities which focus on telling time and on coins and their values.

For each activity described, the mathematical concepts and skills involved are listed, thereby aiding the teacher in finding activities which incorporate the content for which the children have the prerequisite background. Appendix A in this book will also help the teacher find appropriate activities.

Following each description of an activity, instructional uses of the activity are discussed. For example, some activities can be used for initiating a concept; others help children visualize figures and relative values. There are activities during which children discuss and sharpen ideas. Other activities provide needed reinforcement for concepts and skills.

Many of the activities described in this chapter and in previous chapters can be varied to incorporate concepts from other areas of mathematics. An excellent example of an activity which can be adapted for uses with many mathematical topics is *Word Race*, an activity described in this chapter. In fact, it is possible to vary activities sufficiently that motor learning can be included in the instructional program for almost every topic in the mathematics program.

* * * * *

*Concepts*: One- and two-dimensional geometric figures
*Activity*: Show A Shape

Children are scattered about the activity area, far enough apart that each child has space for swinging arms and moving about. The leader calls "Take two turns and show a ____," specifying the two-dimensional geometric figure each child is to form with his arms or body. All children turn around twice, then form the figure named. For example, a circle can be suggested easily with both arms overhead as hands touch. By bending elbows but keeping hands and forearms rigid, different quadrilaterals can be formed. A child who touches his toes while keeping his legs straight makes a triangle. At times have children work in pairs to form figures.

Lines, line segments, rays, and angles can also be shown. Children can let their extended arms represent a part of a line, with a fist used to suggest an endpoint. An infinite extension can be suggested by pointing the finger. For angles, the torso can become the vertex as the arms are swung to different positions. Acute, right, and obtuse angles can all be pictured in this way.

*Instructional Use*: This activity helps children learn that geometric figures are not just marks on paper, that they consist of a set of locations in space. The activity can be used to introduce selected definitions, e.g. an obtuse angle. However, because many of the figures formed will be suggestive rather than precise, the activity will usually be used for reinforcement.

* * * * *

*Concepts*: Triangle
Polygons
*Activity*: Triangle Run

A large triangle is marked off with a base at each vertex. Three teams of equal size are formed, and one team stands behind each base. On a signal the first child of each team leaves his base and runs to his right around the triangle, touching each base on the way. When he returns to his base, the next child on his team does the same. The runners may pass each other, but they must touch each base as they run. The first team back in its original place wins.

*Instructional Use*: This activity helps to demonstrate certain properties of a triangle, for example, the three angles. It is best to mark off different-shaped triangles from time to time so that the properties observed can be generalized to all triangles. Other polygons can be illustrated by forming more than three teams.

* * * * *

*Concept*: Radius of a circle
*Activity*: Jump the Shot

Eight to ten children make a circle, facing the center. One child stands in the center of the circle with a beanbag tied to the end of a rope. The center child swings the rope around in a large circle low to the surface area in order for the beanbag to pass under the feet of those in the circle. The children in the circle attempt to jump over the beanbag as it passes beneath their feet, for when the rope or beanbag touches a child, it is a point against him. The child with the lowest score wins at the end of a period one to two minutes long. The child in the center may then exchange places with a child in the circle.

*Instructional Use*: The activity can be used to help children visualize the radius of a circle. They should note that the rope, which represents the radius, is the same length from the center to any point along the circle.

* * * * *

*Concepts*: A circle is a simple closed curve
The interior of a circle
*Activity*: Run Circle Run

The class forms a circle by holding hands and facing inward. Depending on the size of the group, children count off by two's or three's (for small groups) or four's, five's, or six's (for large groups of around 30). The teacher calls one of the assigned numbers. All the children with that number start running around the circle in a specified direction; each runner tries to tag one or more children running ahead of him. As a successful runner reaches his starting point without being tagged, he stops. Runners who are tagged go to the interior of the circle. Another number is called, and the same procedure is followed. Continue until all have been called, then reform the circle, assign new numbers to the children, and repeat.

As the number of children decreases, a smaller circle can be drawn on the surface area inside the larger circle; the children must stay out of the smaller circle when running around to their places.

*Instructional Use*: Help the children note that when they form a circle by holding hands, they make a continuous, simple, closed shape. As they play the game, they should observe what happens to the size of the circle as parts of it break off. They also learn to call the space within the circle the interior of the circle.

* * * * *

*Concept*: Interior of a figure
*Activity*: Three Bounce Relay

Teams are formed and make rows behind a starting line. A small circle about one foot in diameter is drawn fifteen feet in front of the starting line before each team. At a signal the first child on each team runs with a ball to the circle. At the circle he attempts to bounce the ball three times within the circle, that is, in the interior of the circle. If the ball at any time does not land on the interior, the child must start over from the starting line. When a child has bounced the ball three times within the circle, he returns to the starting line and touches the next child, who does the same thing. The first team finished wins.

*Instructional Use*: The word *interior* is stressed in explaining the game. The children can thus learn the meaning of this term by practical use. However, so that children do not associate the term only with circular regions, other geometric shapes can be drawn from time to time.

* * * * *

*Concept*: Two-dimensional geometric figures
*Activity*: Geometric Figure Relay

Two lines are drawn about thirty feet apart on a playing area. The class is divided into two teams, with both teams standing behind one of the lines. The teacher calls out the name of a geometric figure, and the teams run across to the opposite line and form the figure with hands joined. The first team that forms the figure correctly wins a point. The teams then line up behind that line and, when the teacher calls another figure, they run to the opposite line and again form the figure the teacher has called. The geometric figures can be those that the children have been working with, and may include circle, square, rectangle, triangle, pentagon, rhombus, equilateral triangle, and the like. By letting a raised hand represent an endpoint and an extended hand take the place of an arrow to indicate continuation, two-dimensional figures other than polygons can be called. Possibilities include line, line segment, ray, and half-line.

*Instructional Use*: By acting out their shapes, children gain familiarity with a variety of geometric figures and their properties.

* * * * *

*Concepts*: Geometric figures
Mathematical vocabulary (varied topics)
*Activity*: Word Race

Two teams are selected, and both line up along a base line. Identical sets of cards are prepared for each team, with each card showing a definition or phrase in bold letters. The cards are distributed to members of the two teams with one card (or possibly two) per child. The leader stands beside a box on the playing surface about twenty-five feet from the base line. He has a set of word cards with letters large enough to be read by the children. When the leader holds up a word card, the child from each team with the matching definition or phrase card races to place his card in the box. The first team placing the correct card in the box wins a point. In order to minimize confusion as to which card is placed in the box first, different colors can be used for the two sets of cards.

*Instructional Use*: Children have many opportunities to discuss the definitions of mathematical terms as this activity progresses, and they learn much from each other. In fact, questions or misunderstandings may come to light which the teacher will want to deal with at an appropriate moment. A few examples of words for word cards and of matching definitions or phrases are listed below, but teachers will be able to think of others which tie in directly with mathematics which is the immediate focus of instruction in the classroom. Of course, it is necessary to use definitions appropriate to the child's level of development.

| *Leader's Word Card* | *Matching Card for Team Member* |
| --- | --- |
| triangle | a polygon with three sides |
| rhombus | a polygon with four equal sides |
| interior | includes all the points inside |
| radius | from the center to the circumference |
| addend | tells how many in one of two parts of a set |
| factor | the result of dividing a product by a factor |
| prime number | has exactly two different whole number factors |

* * * * *

*Concepts*: Geometric figures
Mathematical vocabulary (varied topics)
*Activity*: Have You Seen My Friend?

Each child is assigned a mathematical name, a term which has been used in math class. For example, one child may be a triangle and another a cube. Children could also be assigned names like factor, zero, and centimeter. Names should be printed on cards and either pinned on each child or placed around the neck with a string.

Appropriately named, the children stand or sit in a circle. One child is *It*, and he walks around the outside of the circle. Eventually, *It* stops behind one of the players and asks him, "Have you seen my friend?" The child in the circle answers, "What is your friend like?" *It* describes the mathematical concept which is the name of one of the other children in the circle. He may say, "My friend is a polygon, and he has four sides." The child in the circle attempts to guess which other child is being described, and as soon as he guesses correctly, he chases that child around the outside of the circle. The child being chased tries to run around the circle and return to his place without being tagged by the chaser. If he is tagged, he becomes *It.* If he is not tagged, the chaser becomes *It.* The child who was *It* before the chase merely steps into the place made vacant by the chaser.

*Instructional Use*: During this activity, discussions concerning correct and adequate definitions of mathematical concepts are likely to evolve. For example, if a child describes his friend as "a polygon with four sides" and one child is named "square" and another "rhombus," the fact that the description could have applied to either is likely to come to light. When selecting mathematical terms to use, needed reinforcement can be provided by focusing on ideas which have been studied recently.

* * * * *

*Concept*: Parallel lines
Right angles
*Activity*: Streets and Alleys

The children divide into three or more parallel lines with at least three feet between children in each direction. A runner and a chaser are chosen. The children all face the same direction and join hands with those on each side forming "streets" between the rows. Dropping hands, the children make a quarter turn and join hands again and form "alleys." The chaser tries to tag the runner going up and down the streets or alleys but not breaking through or going under arms. During the game, the teacher aids the runner by calling "streets" or "alleys" at the proper time. At this command the children drop hands, turn, and grasp hands in the other direction, thus blocking the passage for the chaser. When caught, the runner and chaser select two others to take their places.

*Instructional Use*: Children should be helped to note that the lines of children represent parallel lines. Further, when the teacher calls "alleys" and the children make a quarter turn, children can associate the turn with a right angle.

* * * * *

*Concepts*: Circle
Inside and outside
Exploration of space
*Activity*: Inside Out

The class is divided into teams of four or more, and children on each team join hands to form a circle in which each child is facing toward the inside. When the leader calls "Inside out," each team tries to turn its circle inside out. That is, while *continuing* to hold hands the children move so as to face out instead of in. To do this, a child will have to lead his team under the joined hands of two team members. The first team to complete a circle with children facing toward the outside of the circle wins.

*Instructional Use*: This activity is designed as a kind of puzzle or problem-solving activity, for the goal is presented to the children and they are not told how it is possible to turn the circle inside out. It is designed to be the initial encounter with the process involved.

* * * * *

*Concepts*: Three-dimensional geometric figures
Simple logic
*Activity*: Three-D Race

The class is divided into teams which line up behind a starting line. A finish line is marked about fifty feet from the starting line and parallel to it. At the side of the first runner for each team is the team's collection of models of three-dimensional geometric shapes. The models can be purchased commercially (probably made of wood or plastic), or made from boxes, cans, and the like. The collections, which should be alike for all teams, are placed on the surface area within a tray, box lid, or a similar container. The teacher calls out the name for a shape or properties identifying a class of shapes, then the first runner selects one shape which fits the identification and runs to the finish line. The first runner to reach the finish line with a correctly selected shape scores one point for his team.

The teacher can make many different calls depending upon the experience the children have had with shapes and their properties. Possible calls include: cube, pyramid, prism, cylinder, cone, triangular pyramid, a shape with twelve edges, a prism but not a cube, a shape with exactly six faces, a figure with both curved and plane surfaces, an ellipsoid, and so forth.

*Instructional Use*: While engaged in this activity, children observe properties of different three-dimensional shapes and are helped to associate names with the major classes of shapes. A review of the different shapes and how they are identified would be appropriate before the game is played. An emphasis on logic can be incorporated by introducing additional variables (such as color) and negations. For example, the teacher could call "Red and not a rectangular prism."

* * * * *

*Concept*: Three-dimensional geometric figures
*Activity*: Pass the Shape

Children are divided into groups of about ten players. In each group, one child is *It* and the others sit close together in a small circle about thirty feet from a safety line; the child who is *It* stands in the center of the circle. Each child in a circle holds behind his back a small model of one of the three-dimensional shapes which the children have learned about: cubes, triangular pyramids, spheres, cones, and the like. They may be made of wood, plastic, or any durable material. The child who is *It* calls "Pass the shape," then spins completely around five times counting the spins as he does so. As soon as *It* calls "Pass the shape," the children in the circle begin passing the models around the circle clockwise, keeping them behind their backs and taking care to keep them out of sight of *It*. As *It* finishes his five spins he says "Stop," at which time the children stop passing the shapes. *It* then points to one child and asks, "Do you have a ____?" (naming the shape he thinks the child is holding). The child in the circle holds up his shape for all to see. If *It* does not guess correctly, he calls "Pass the shape" and spins again. If *It* does guess correctly, he goes to a place half-way between the circle of children and the safety line and calls "Run for safety." The children all run to the safety line as *It* tries to tag as many children as possible. The first child tagged is the new *It* when the children form a new circle and play continues.

*Instructional Use*: As children are involved in this activity, they focus on properties of three-dimensional figures: smooth flat surfaces, edges, corners, and so forth. When discussing the shapes, more technical vocabulary can be introduced gradually and as it meets a need for more precise communication. It may or may not be desirable to allow use of words which name very general categories, for example, a polyhedron.

The models of three-dimensional shapes need not be commercially prepared models, though these are very satisfactory. The use of everyday objects for this purpose will help children see mathematics in the world about them. Examples include typical building blocks (cube and rectangular prism), a marble (sphere), a small jar or can (cylinder), a smooth stone (ovoid), a small box (rectangular prism), a penny (cylinder), and Cuisenaire rods (cube and rectangular prism).

The difficulty of the activity depends somewhat upon the number of different shapes used at any one time. Each child in a group can be given a different shape; or, to make the activity easier, several children can be given the same shape.

* * * * *

*Concept*: Recognition of coins
*Activity*: Name the Coin

The class is divided into several teams, and a box containing a quantity of pennies, nickels, dimes, quarters, and possibly half dollars is given to each team. A leader from each team is assigned the task of keeping track of the team's money box. The teams stand in rows behind a line a specified distance from the leaders. The teacher calls out the name of any coin, then the first member of each team runs to his team's money box, finds the coin in the box, shows it to the teacher, replaces it, and returns to the team. The child who first identifies the correct coin in the team box scores a point for his team. This continues, the teacher naming different coins for members of the teams to find and identify. The team with the most points wins.

*Instructional Use*: This game, which provides reinforcement for recognition of coins, would be played after introduction of these coins and their values. A short review preceding the game will help children to recall the names and values of the coins.

* * * * *

*Concept*: Recognition of coins
*Activity*: Stepping Stones

On the activity area draw a stream with stepping stones arranged so that a child can take different paths across the stream. Place a coin on each stepping stone. Caution the children not to fall in the stream by making a mistake naming the coins. As a child chooses a stepping stone to cross the stream and steps on the stone, he must name the coin and its value. Children at their seats watch for mistakes. A successful crossing scores one point for the child. It is important to change the coins frequently.

*Instructional Use*: This game provides an interesting context for children to practice naming coins and stating their values. As children watch they too are involved in the game, confirming and/or rejecting their own ideas about the coins.

* * * * *

*Concept*: Values of coins
*Activity*: Bank

One child is selected to be the Bank. The rest of the children stand at a starting line about twenty feet from the Bank. Bank calls out the number of pennies, nickels, or dimes a child may take. (A penny is one small step; a nickel equals five penny steps; and a dime equals ten penny steps.) Bank calls a child's name and says, "Harry, take three pennies." Harry must answer "May I?" Then Bank can say "Yes, you may" or "No, you may not." If Harry forgets to say "May I?" he must return to the starting line. The first child to reach the Bank becomes the Bank. All children should be called upon.

*Instructional Use*: By taking the steps forward toward the Bank, children gain an understanding that a nickel has the same value as five pennies and a dime has the same value as ten pennies. The teacher may want to encourage each child to say before moving forward how many steps he is allowed to take.

* * * * *

*Concept*: Values of coins
*Activity*: Banker and the Coins

Each child in the class is given a sign to wear which denotes one of the following: five cents, ten cents, twenty-five cents, fifty cents, nickel, dime, quarter, or half-dollar. One child is the Banker, who calls out different amounts of money up to one dollar. The children run and group themselves with other children until their group totals the amount of money called by the Banker. Every child who is part of a correct group gets a point.

*Instructional Use*: Children gain practice in combining different amounts of money to arrive at a specific amount. They learn that there are usually several different ways one can combine coins to produce a specified amount of money. As the teacher checks that a group is correct, he can have the whole class count it out. In this way, children will more likely learn about values of coins if they are unsure.

* * * * *

*Concept*: Telling time
*Activity*: Tick Tock

The class forms a circle that represents a clock. Two children are runners and they are called Hour and Minute. The children chant "What time is it?" Minute then chooses the hour and calls it out (six o'clock). Hour and Minute must stand still while the children in the circle call "one o'clock, two o'clock, three o'clock ... six o'clock" (or whatever time has been chosen). When the children get to the chosen hour, the chase begins. Hour chases Minute clockwise around the outside of the circle. If Hour can catch Minute before the children in the circle once again call out "one o'clock, two o'clock ... six o'clock" (the same hour as counted the first time), he chooses another child to become Hour. The game can also be played counting by half-hours.

*Instructional Use*: Children not only get practice in calling the hours in order but also gain experience with the concept of the term *clockwise*.

* * * * *

*Concept*: Telling time
*Activity*: Set the Clock Relay

Two clock faces are drawn on pieces of heavy tagboard and movable hands are attached to each clock face. The clocks are then secured to a bulletin board or to a chalkboard. Children form two teams in relay formation about ten feet from the clock faces. The teacher calls out a time such as "seven o'clock." The first child on each team runs, skips, or hops (as directed) to his team's clock and places the hands at seven o'clock. The teacher calls another time, and the second child on each team sets the hands of the clock for that particular time. Each child proceeds until everyone on each team has had a chance to set the clock. For each time called out, the child who sets the clock first and accurately scores a point for his team. The team with the most points wins.

*Instructional Use*: This relay provides needed practice for developing skill in telling time. Children have their own turn at setting the clock and they can also watch the other children to be sure they are not making mistakes. If even hours are being called, place clock hands at random after each play so children will get practice deciding which hand to place at the 12.

* * * * *

*Concept*: Telling time
*Activity*: Clock Race

The class is divided into teams as in other relays. The teacher prepares slips of paper containing statements related to time, such as "Recess begins at 10:40," "The buses leave at 3:15," or "The Flintstone Show begins at 4:00." These statements are placed in a small container in front of each team, and clock faces with movable hands are placed on the chalkboard ledge. Captains are appointed for each team, and at a given signal the captains each draw a statement from the container and read it. The first child on each team runs and sets the team's clock to the specified time. The child then returns to the team. The captain reads another statement, and the second child on the team sets the team's clock accordingly. The game continues until one team makes all the clock settings as required. However, if an error occurs, the captain is responsible for detecting it and must correct it before his team can proceed to the next statement. The teacher helps to keep track of any errors. The first team finished wins.

*Instructional Use*: Through this interesting activity, time-telling skills are reinforced. The game can be easily adapted to different levels of difficulty for different groups of children.

* * * * *

*Concept*: Telling time
*Activity*: Five-Minute Relay

The class is divided into two teams. A cardboard clock with movable hands is set up for each team at the front of the room. Each clock is set at any given time, and may be set at different times so the teams cannot copy each other. The teams line up in relay formation a specified distance from the clocks. On a signal the last child in each team's line begins the game by running forward to his team's clock. He moves the minute hand ahead by five minutes and writes down the new time on the board. Meanwhile, his team members have moved back so that as the child returns from the clock, he can return to the front of the team's line. As soon as the first runner returns, he raises his hand, and the last child in the line proceeds. He moves the minute hand ahead five minutes and writes down the time. The first team whose runners all finish setting the clock and writing down the time correctly wins.

*Instructional Use*: In this activity, children are provided the repetition necessary for learning to read a clock and record time quickly and accurately. In situations that are highly motivating, the game reinforces skills previously presented.

* * * * *

*Concept*: Linear measurement
*Activity*: Ring Toss

A regular ring toss game can be used for this activity, and a meter stick or a yardstick is also required. The class is divided into two teams, and each team has a ring. One post is used for both teams. The first child on each team takes a turn tossing his ring, then each takes the measuring stick and measures the distance between the post and his ring. The child whose ring is closest to the post scores one point for his team. Before beginning the game, the teacher should indicate how precisely measurements are to be made. For example, if the meter stick is used, measurements will probably be to the nearest centimeter. If a yardstick is used, the nearest quarter-inch might be specified. The team with the highest score wins.

*Instructional Use*: This game provides each child a chance to develop his measuring skills. He will get additional practice if he checks the opposing team member's ring.

* * * * *

*Concepts*: Linear measurement
Average (mean)
Addition (including decimals)
*Activity*: Add-A-Jump Relay

The class is divided into teams, and each team has one child serving as marker and one child serving as scorekeeper. Teams line up in a single-line formation. The first child on each team moves up to the starting line in front of his team and jumps as far as he can. The team's marker indicates where the child landed, then the child who jumped measures the distance he jumped. The team's scorekeeper records the distance, and the second child on the team moves to the starting line and jumps. The relay continues until all team members have jumped. Scorekeepers add the measurements, and the team with the greatest total distance is the winner.

*Instructional Use*: In addition to developing measuring skills further, this activity provides practice in addition and experience comparing distances. Either English or metric units can be used for measuring. More advanced children could measure in meters and express the distance with a decimal including tenths and hundredths (to the nearest centimeter). When the children return to the classroom, they can use the figures that were recorded to find the average distance jumped and to compare team or individual records with previously established records.

* * * * *

*Concepts*: Average (mean)
Linear measurement
Addition and division
*Activity*: Beanbag Throw

Two parallel lines five to ten feet in length are marked approximately fifty to eighty feet apart, depending on the children. The children are divided into two groups; each group is assigned to a line. One child from the first group is given a beanbag, and he begins the game by tossing it from behind his starting line. One child from the opposite side then marks the spot where the beanbag landed, while another child measures the distance the beanbag was thrown with a tape measure. A scorekeeper records the distance. A child in the second group is now allowed to throw a beanbag from the starting line for his group; it is marked, measured, and recorded by members of the first group. Play continues until all children have had three throws, and the child with the longest average distance is the winner.

*Instructional Use*: By doing their own computing, children learn what an average is and how to find an average; therefore, it is best to have each child compute his own average. This activity also provides practice in measuring, adding, and dividing.

* * * * *

*Concepts*: Linear measurement
Meters, decimeters, and centimeters
*Activity*: Metric Race

Children are divided into two teams, and parallel metric number scales are drawn on the playing surface, one scale for each team. Start each scale with zero and extend 8-10 meters or more. Markings should include at least meters and decimeters.

Each team lines up at the zero end of a scale, with the first child standing on zero. The leader then calls out the distance to be run. For example, he could call "The number of meters to run is eight," or "The number of decimeters to run is 75." The children standing on zero race for the distance specified, then stoop and touch the mark on the scale that shows the distance from zero which was called. For example, if the call was "The number of decimeters to run is 59," children would run about six meters and point to the mark one decimeter short of the six meter mark. The first child to touch the correct mark scores a point for his team. The next child in each line then stands on zero and another distance is called. The race proceeds similarly until every child has raced once. The team with the highest score wins.

*Instructional Use*: As they participate in this activity, children acquire a feeling for selected metric units. They come to know about how long a meter is, and they become better judges of about how far five meters is. Further, they are enabled to visualize the relative sizes of different metric units. They also reinforce the skills involved in translating measurements *within* the metric system. For example, the call "The number of decimeters to run is 70" requires running to the seven meter mark.

Centimeters can also be incorporated if the scales are marked on tape attached to the floor. Different colors of felt-tipped pens can be used in marking these more detailed scales. An appropriate call might be, "The number of centimeters to run is 675."

* * * * *

*Concept*: Liquid measurement (pints and quarts)
*Activity*: Milkman Tag

Two teams of three milkmen are selected and given milk truck bases. One team might be called Chocolate, the other White. The remaining children are called Pints. On a given signal one milkman from each team tries to tag any one of the Pints. When he tags one, they both go to the milk truck, and another milkman goes after a Pint. The teacher sets a goal of so many quarts to be gained in order to win. The children must then figure out how many Pints will be needed to make the necessary number of quarts.

*Instructional Use*: Children are provided a highly motivating activity for applying the idea that two pints are equivalent to one quart. Children may want to group Pints in two's in order to determine how many quarts each team has.

Appendix A

# LIST OF ACTIVITIES BY CONCEPT


| *Concept* | *Activity* | *Page* |
|---|---|---|
| Addend pairs for a given sum | Count the Pins | 91 |
| | Addition Name Hunt | 106 |
| | Fast Facts | 107 |
| | Parts of Seven | 108 |
| | Who Am I? | 116 |
| Addition, whole numbers | Card Toss | 88 |
| | Addition Bean bag Throw | 109 |
| | Add-A-Number Relay | 110 |
| | Add-A-Jump Relay | 144 |
| | Beanbag Throw | 145 |
| Basic facts (*Also see* Basic facts of arithmetic) | Exchange Numbers | 46 |
| | Fast Facts | 107 |
| | Beanbag Toss | 108 |
| | Number Catch | 110 |
| | Steal the Bacon (Variation) | 111 |
| | Trade Places Tag | 112 |
| | Number Man (Variation) | 112 |
| Column | Count the Pins | 91 |
| Commutativity | Move Like Animals | 76 |
| Higher decade | Beanbag Toss | 108 |
| | Who Am I? | 116 |
| Meaning of | Lions and Hunters | 106 |
| Multidigit numbers | Bean Bag Addition | 109 |
| Average (mean) | Add-A-Jump Relay | 144 |
| | Beanbag Throw | 145 |
| Basic facts of arithmetic | Club Snatch | 39 |
| | Call Ball (Variation) | 123 |
| | Cheese and Mice | 124 |
| | Arithmetic Relay | 124 |
| | Catch the Thief | 125 |
| Circle | Run Circle Run | 132 |
| | Inside Out | 137 |
| Coins | | |
| Recognition of | Name the Coin | 139 |
| | Stepping Stones | 140 |
| Values of | Bank | 140 |
| | Banker and the Coins | 141 |
| Common factors | Catch the Thief | 125 |
| Common multiples | Multiple Squat | 120 |

| *Concept* | *Activity* | *Page* |
|---|---|---|
| Counting | | |
| By tens | | |
| | Bouncing Relay | 84 |
| | Red Light | 96 |
| By twos | | |
| | Find a Friend | 71 |
| Rational (0-9) | | |
| | Hot Spot | 85 |
| | Watch the Numerals | 85 |
| | Show-Me Relay | 90 |
| Rational | | |
| | Mrs. Brown's Mouse Trap | 89 |
| | Move Like Animals | 76 |
| | Bee Sting | 86 |
| | Chain Tag | 86 |
| | Round Up | 87 |
| | Fish Net | 87 |
| | Card Toss | 88 |
| | Ball Bounce | 88 |
| | Ball Pass | 89 |
| | Count the Pins | 91 |
| | Count and Go | 92 |
| Rote | | |
| | Ten Little Indians | 40 |
| | Pass Ball Relay | 83 |
| | Catch A Bird Alive | 83 |
| | Bouncing Relay | 84 |
| Decimals | | |
| | Roving Decimal Point | 103 |
| Addition of | | |
| | Dash Relay | 128 |
| | Add-A-Jump Relay | 144 |
| Divisibility rules | | |
| | Factor First | 120 |
| Division, whole numbers | | |
| | Beanbag Throw | 145 |
| Basic facts (See *Basic facts of arithmetic*) | | |
| Meaning of (measurement) | | |
| | Get Together | 55 |
| | Triplet Tag | 121 |
| | Birds Fly South | 121 |
| | Birds Fly South (Variation) | 122 |
| One-digit divisors | | |
| | Factor First | 120 |
| Ten as divisor | | |
| | Catch the Cane | 49 |
| Divisor, effect of increasing or decreasing | | |
| | Birds Fly South | 121 |
| | Birds Fly South (Variation) | 122 |
| Doubling | | |
| | Muffin Man | 95 |
| | Grand March | 117 |
| Equals | | |
| | Lions and Hunters | 106 |

| *Concept* | *Activity* | *Page* |
|---|---|---|
| Even and odd numbers | Number Man | 92 |
| Exterior (outside) of a figure | Inside Out | 137 |
| Factors, whole number | Factor First | 120 |
| | Catch the Thief | 125 |
| Fractions | | |
| Addition (common denominators) | Hit the Target | 126 |
| Addition (unlike denominators) | Hit the Target (Variation) | 127 |
| Equivalent | Fraction Target Relay | 101 |
| Meaning of (part of a set) | Squat Thrust | 42 |
| | Pop Goes the Weasel | 43 |
| | Corner Spry | 101 |
| | End Ball | 102 |
| | Hit or Miss | 103 |
| Meaning of (part of a unit region) | Flannelboard Fractions | 100 |
| | Fraction Target Relay | 101 |
| Multiplication | Fraction Race | 128 |
| Subtraction (common denominators) | Train Dodge | 127 |
| Geometric figures | Word Race | 134 |
| | Have You Seen My Friend? | 135 |
| One dimensional | Show A Shape | 131 |
| Two dimensional | Show A Shape | 131 |
| | Geometric Figure Relay | 133 |
| Three dimensional | 3-D Race | 137 |
| | Pass the Shape | 138 |
| Greater than | Lions and Tigers | 38 |
| | Bee String | 86 |
| | Mrs. Brown's Mouse Trap | 89 |
| | Number Man | 92 |
| | Catch the Thief | 125 |
| Greatest common factors | Catch the Thief | 125 |
| Interior (inside) | | |
| Of a circle | Run Circle Run | 132 |
| Of a figure | Three-Bounce Relay | 133 |
| | Inside Out | 137 |
| Less than | Lions and Tigers | 38 |

| *Concept* | *Activity* | *Page* |
|---|---|---|
| | Bee Sting | 86 |
| | Mrs. Brown's Mouse Trap | 89 |
| | Number Man | 92 |
| | Ten Little Birds | 113 |
| | Dodge Ball | 114 |
| | Musical Chairs | 114 |
| | Ten Little Chickadees | 115 |
| | Catch the Thief | 125 |
| Logic | | |
| | 3-D Race | 137 |
| Measurement | | |
| Linear | | |
| | Ring Toss | 143 |
| | Add-A-Jump Relay | 144 |
| | Beanbag Throw | 145 |
| | Metric Race | 145 |
| Liquid | | |
| | Milkman Tag | 146 |
| Metric | | |
| | Metric Race | 145 |
| Missing addends | | |
| | Count the Pins | 91 |
| | Parts of Seven | 108 |
| | Who Am I? | 116 |
| Multiples | | |
| | Call and Catch | 94 |
| | Roman Numeral Relay | 96 |
| | Number Line Relay | 98 |
| | Back to Back | 119 |
| | Multiple Squat | 120 |
| | Catch the Thief | 125 |
| Of two | | |
| | Twice As Many | 117 |
| Of three | | |
| | Pick Up Race | 118 |
| | Leader Ball | 93 |
| Multiplication, whole numbers (Also see *Basic facts of arithmetic*, and *multiples*) | | |
| | Back to Back | 119 |
| By powers of ten | | |
| | Catch the Cane | 49 |
| Number | | |
| Cardinal | | |
| | Ten Little Indians | 40 |
| | Move Like Animals | 76 |
| | Bouncing Relay | 84 |
| | Roman Numeral Bounce | 97 |
| Ordinal | | |
| | Call and Catch | 94 |
| | Ten Little Indians | 40 |
| | Pass Ball Relay | 83 |
| | Bouncing Relay | 84 |
| | Number Man | 92 |
| | Leader Ball | 93 |
| | Number Race | 93 |

| *Concept* | *Activity* | *Page* |
|---|---|---|
| | Fetch and Carry | 94 |
| | Roman Numeral Relay | 96 |
| Number line | | |
| | Number Line Relay | 98 |
| Number sentences | | |
| | Lions and Hunters | 106 |
| Number words (0-9) | | |
| | Show-Me Relay | 90 |
| Numeration | | |
| | Pass Ball Relay | 83 |
| | Hot Spot (0-9) | 85 |
| | Watch the Numerals (0-9) | 85 |
| | Show-Me-Relay (0-9) | 90 |
| | Muffin Man | 95 |
| | Number Line Relay | 98 |
| | Postman Game | 98 |
| | Find Your Place | 99 |
| | Place Value Relay | 99 |
| Reading of numerals | | |
| | Ball Bounce (0-9) | 88 |
| | Show-Me-Relay (0-9) | 90 |
| | Count and Go | 92 |
| | Postman Game | 98 |
| Odd and even numbers | | |
| | Number Man | 92 |
| One | | |
| Factor of every whole number | | |
| | Factor First | 120 |
| Named by n/n | | |
| | Triple Change | 50 |
| Parallel lines | | |
| | Streets and Alleys | 136 |
| Polygons | | |
| | Triangle Run | 131 |
| Powers of two | | |
| | Muffin Man | 95 |
| Prime Numbers | | |
| | Catch the Thief | 125 |
| Radius | | |
| | Jump the Shot | 132 |
| Right angles | | |
| | Streets and Alleys | 136 |
| Roman numerals | | |
| | Roman Numeral Relay | 96 |
| | Roman Numeral Bounce | 97 |
| Space, exploration of | | |
| | Inside Out | 137 |
| Subtraction, whole numbers | | |
| | Add-A-Number Relay | 110 |
| Basic facts (*Also see* Basic facts of arithmetic) | | |
| | Parts of Seven | 108 |
| | Number Catch | 110 |
| | Steal the Bacon (Variation) | 111 |
| | Trade Places Tag | 112 |

| *Concept* | *Activity* | *Page* |
|---|---|---|
| | Number Man (Variation) | 112 |
| | Who Am I? | 116 |
| Meaning of | | |
| | Ten Little Birds | 113 |
| | Dodge Ball | 114 |
| | Ten Little Chickadees | 115 |
| | Tossing Darts | 116 |
| Time, telling | | |
| | Tick Tock | 141 |
| | Set the Clock Relay | 142 |
| | Clock Race | 142 |
| | Five Minute Relay | 143 |
| Triangle | | |
| | Triangle Run | 131 |
| Vocabulary, varied mathematical topics | | |
| | Word Race | 134 |
| | Have You Seen My Friend? | 135 |
| Zero, number of the empty set | | |
| | Ball Bounce | 88 |
| | Show-Me Relay | 90 |

## Appendix B

# LIST OF ACTIVITIES BY TITLE


| *Activity* | *Page* |
|---|---|
| Add-A-Jump Relay | 144 |
| Add-A-Number Relay | 110 |
| Addition Beanbag Throw | 109 |
| Addition Name Hunt | 106 |
| Arithmetic Relay | 124 |
| | |
| Back to Back | 119 |
| Ball Bounce | 88 |
| Ball Pass | 89 |
| Bank | 140 |
| Banker and the Coins | 141 |
| Bean Bag Addition | 109 |
| Beanbag Throw | 145 |
| Beanbag Toss | 108 |
| Bee Sting | 86 |
| Birds Fly South | 121 |
| Birds Fly South (Variation) | 121 |
| Bouncing Relay | 84 |
| | |
| Call and Catch | 94 |
| Call Ball (Variation) | 123 |
| Card Toss | 88 |
| Catch A Bird Alive | 83 |
| Catch the Cane | 49 |
| Catch the Thief | 125 |
| Chain Tag | 86 |
| Cheese and Mice | 124 |
| Clock Race | 142 |
| Club Snatch | 39 |
| Corner Spry | 101 |
| Count and Go | 92 |
| Count the Pins | 91 |
| | |
| Dash Relay | 128 |
| Dodge Ball | 114 |
| | |
| End Ball | 102 |
| Exchange Numbers | 46 |
| | |
| Factor First | 120 |
| Fast Facts | 107 |
| Fetch and Carry | 94 |
| Find A Friend | 71 |
| Find Your Place | 99 |
| Fish Net | 87 |
| Five-Minute Relay | 143 |
| Flannelboard Fractions | 100 |
| Fraction Race | 128 |
| Fraction Target Relay | 101 |

| *Activity* | *Page* |
|---|---|
| Geometric Figure Relay | 133 |
| Get Together | 55 |
| Grand March | 117 |
| Have You Seen My Friend? | 135 |
| Hit or Miss | 103 |
| Hit the Target | 126 |
| Hit the Target (Variation) | 127 |
| Hot Spot | 85 |
| Inside Out | 137 |
| Jump the Shot | 132 |
| Leader Ball | 93 |
| Lions and Hunters | 106 |
| Lions and Tigers | 38 |
| Metric Race | 145 |
| Milkman Tag | 146 |
| Move Like Animals | 76 |
| Mrs. Brown's Mouse Trap | 89 |
| Muffin Man | 95 |
| Multiple Squat | 120 |
| Musical Chairs | 114 |
| Name the Coin | 139 |
| Number Catch | 110 |
| Number Line Relay | 98 |
| Number Man | 92 |
| Number Man (Variation) | 112 |
| Number Race | 93 |
| Parts of Seven | 108 |
| Pass Ball Relay | 83 |
| Pass the Shape | 138 |
| Pick-Up Race | 118 |
| Place Value Relay | 99 |
| Pop Goes the Weasel | 43 |
| Postman Game | 98 |
| Red Light | 96 |
| Ring Toss | 143 |
| Roman Numeral Bounce | 97 |
| Roman Numeral Relay | 96 |
| Round Up | 87 |
| Roving Decimal Point | 103 |
| Run Circle Run | 132 |
| Set the Clock Relay | 142 |
| Show A Shape | 131 |
| Show-Me Relay | 90 |
| Squat Thrust | 42 |
| Steal the Bacon (Variation) | 111 |
| Stepping Stones | 140 |
| Streets and Alleys | 136 |

| *Activity* | *Page* |
|---|---|
| Ten Little Birds | 113 |
| Ten Little Chickadees | 115 |
| Ten Little Indians | 40 |
| Three-Bounce Relay | 133 |
| 3-D Race | 137 |
| Tick Tock | 141 |
| Tossing Darts | 116 |
| Trade Places Tag | 112 |
| Train Dodge | 127 |
| Triangle Run | 131 |
| Triple Change | 50 |
| Triplet Tag | 121 |
| Twice As Many | 117 |
| Watch the Numerals | 85 |
| Who Am I? | 116 |
| Word Race | 134 |

# INDEX


**A**

Ammons, R. B., 32
Aristotle, 53
Armington, John C., 33
Asher, James J., 23
Ashlock, Robert B., 16

**B**

Biggs, Edith E., 15
Biggs, John, 15
Bilodeau, Edward A., 32
Bilodeau, Ina, 32
Brown, J. S., 32
Brownell, William A., 10

**C**

Cilley, David M., 14
Comenius, 53
Contemporary mathematics programs, 11-15
 characteristics, 12-13
 contributing factors, 11-12
 criticism of, 13
 laboratory approach, 14-15
 structural apparatus, 13-14
Crist, Thomas, 59

**D**

Dalrymple, Charles O., 6
Deans, Edwina, 12
DeVault, M. Vere, 3, 5, 6
Dewey, John, 36, 54
Droter, Robert, 61

**F**

Fish, Daniel W., 5, 6
Froebel, Friedrich, 53

**G**

Games for learning, 15-19
 for the slow learner, 17-19
Glennon, Vincent J., 13
Good, Carter V., 20
Gunderson, Agnes G., 83

**H**

Hall, G. Stanley, 54
Historical background of mathematics teaching, 3-7
 colonial period, 3-4
 nineteenth century, 4-6
 twentieth century, 6-7
Hollister, George E., 83
Humphrey, James, 21, 24, 58, 68, 70, 73, 74, 87, 89, 94
Hunter, Edward, 36

**I**

Ivanitchkii, M. F., 37

**J**

Jacks, L. P., 54
Johnson, G. B., 22
Johnson, W. R., 22
Junge, Charlotte W., 18

**K**

Kennedy, Leonard M., 16
Kidd, Kenneth P., 14
Kinney, Lucien B., 18
Kriewall, Thomas E., 3
Krug, Frank, 63

## L


Lee, Dorris May, 31
Lee, J. Murray, 32


## M


MacGinnis, George H., 78
MacLean, James R., 15
McKenzie, William G., 17
Marks, John L., 18
Mathematics motor activity stories, 68-81
  experiments with, 70-74
    materials resulting from, 74-77
  guidelines for use of, 79-81
  prepared by teachers, 77-79
Michon, Ruth L., 16
Monroe, Walter S., 3
Montessori, Maria, 13
Moore, Virginia D., 68, 70
Motor activity, 20-37
  current status and future prospects of learning through, 35-36
  factors influencing learning through, 27-35
    motivation, 30-33
      in relation to competition, 32-33
      in relation to interest, 31
      in relation to knowledge of results, 31-32
    proprioception, 33-34
    reinforcement, 34-35
  general ways of providing mathematics experiences through, 38-52
    game activities, 38-40
    rhythmic activities, 40-41
    self-testing activities, 41-43
    simulated teaching-learning situations, 43-52
  learning about numbers and numeration systems through, 82-104
  learning about operations of arithmetic through, 105-129
  nature of learning through, 20-37
  theory of learning through, 24-27
Motor learning, 20-27
  branches of, 21-27
    involving academic skill and concept development, 23-24
    involving learning of motor skills, 21-22
    involving perceptual-motor development, 22-23
  meaning of, 20
  research in learning about mathematics through, 53-67
    representative findings, 58-66
      generalizations, 66-67
    techniques, 54-58
      naturalistic observation, 55-56
      parallel groups, 56-57
      single groups, 56
      variations of standard procedures, 57-58
Myers, Shirley S., 14


## P


Paschal, Billy J., 18
Pearson, James R., 17, 99
Piaget, Jean, 10
Plato, 53
Post, Thomas R., 14
Purdy, C. Richard, 18


## Q


Quintillian, 53


## R


Riedesel, C. Alan, 5
Reissman, Frank, 18
Reys, Robert E., 14
Rousseau, J. J., 53
Rowan, Thomas E., 17


## S


Schultz, Richard W., 18
Spache, George D., 68, 78
Spencer, Herbert, 54
Steinhaus, Arthur H., 34
Stone, Mildred B., 6


## T


Theories of mathematics teaching, 7-11

Gestalt, 9-10
Meaning, 10-11
Social Utility, 8-9
Stimulus-response, 7-8
Thorndike, Edward L., 8
Trout, Edwin, 65

**W**

Wilson, Guy M., 6
Wright, Charles, 62

**Y**

Yessis, Michael, 37